CRISIS AND REFLECTION

PHAENOMENOLOGICA
SERIES FOUNDED BY H.L. VAN BREDA AND PUBLISHED
UNDER THE AUSPICES OF THE HUSSERL-ARCHIVES

174

JAMES DODD

CRISIS AND REFLECTION

An Essay on Husserl's *Crisis of the European Sciences*

JAMES DODD
Graduate Faculty, New School University

CRISIS AND REFLECTION

An Essay on Husserl's
Crisis of the European Sciences

KLUWER ACADEMIC PUBLISHERS
DORDRECHT / BOSTON / LONDON

A C.I.P. Catalogue record for this book is available from the Library of Congress.

ISBN 978-90-481-6606-0 (PB)
ISBN 978-1-4020-2175-6 (e-book)

Published by Kluwer Academic Publishers,
P.O. Box 17, 3300 AA Dordrecht, The Netherlands.

Sold and distributed in North, Central and South America
by Kluwer Academic Publishers,
101 Philip Drive, Norwell, MA 02061, U.S.A.

In all other countries, sold and distributed
by Kluwer Academic Publishers,
P.O. Box 322, 3300 AH Dordrecht, The Netherlands.

Printed on acid-free paper

Printed in the Netherlands.

Table of Contents

Abbreviations of Husserl's Works

Published Works

With the exception of *Erfahrung und Urteil*, all citations from Husserl's published works are from Husserliana, and are given in the form: Hua + volume + page:line. Current English translations of Husserl's works have been cited from, whenever possible (see Bibliography for translations used). If the work is unavailable in translation, I have provided translations when the quote appears in the main body of the text. If the text is cited in the original language edition alone, the translation is my own. The following abbreviations of Husserl's works have been employed:

EU	Erfahrung und Urteil
EJ	Experience and Judgment
Hua III,1	Ideen zu einer reinen Phänomenologie und phänomenologischen Philosophie. Erstes Buch
Ideas I	Ideas Pertaining to a Pure Phenomenology and a Phenomenological Philosophy. First Book
Hua IV	Ideen zu einer reinen Phänomenologie und phänomenologischen Philosophie. Zweites Buch
Ideas II	Ideas Pertaining to a Pure Phenomenology and a Phenomenological Philosophy. Second Book
Hua VI	Die Krisis der europäischen Wissenschaften und transzendentale Phänomenologie
Crisis	The Crisis of European Sciences and Transcendental Phenomenology
Hua VII	Erste Philosophie (1923-1924). Erster Teil: Kritische Ideengeschichte
Hua VIII	Erste Philosophie (1923-1924) Zweiter Teil: Theorie der phänomenologischen Reduktion
Hua X	Zur Phänomenologie des inneren Zeitbewusstseins (1893-1917)

ITC	On the Phenomenology of the Consciousness of Internal Time (1893-1917)
Hua XI	Analysen zur passiven Synthesis
Hua XVII	Formale und transzendentale Logik
FTL	Formal and Transcendental Logic
Hua XVIII	Logische Untersuchungen. Erster Band: Prolegomena zur reinen Logik
LU,1	Logical Investigations. Volume One
Hua XIX,1/2	Logische Untersuchungen. Zweiter Band. Teil I/II
LU,2	Logical Investigations. Volume Two
Hua XXVII	Aufsätze und Vorträge (1922-1937)
Hua XXIX	Die Krisis der europäischen Wissenschaften. Ergänzungsband
Hua XXXI	Aktive Synthesen. Ergänzungsband zu "Analysen zur passiven Synthesis"

Unpublished Manuscripts

The following manuscripts are cited in accordance with the index used at the Husserl-Archiv in Cologne and the Husserl-Archief in Leuven. Page numbers refer to the original manuscript.

C 3 III (March 1931) "…—Strukturen der lebendigen Gegenwart: passive Association, Aktstruktur, Wachheit und Unwachheit, Aktmodi […] Ständigkeit der Konstitution, Sedimentierung und Weckung; Schichten und Stufen der Zeitigung"

C 6 (August 1930) "Erster Anfang eines methodischen Abbaus der urphänomenalen Gegenwart …"

C 16 I (Ende 1931) "Fundamentalanalyse der lebendigen Gegenwart nach der Aktstruktur; das Wache (Patente) und Latente (Unbewusste); die Aktmodi; auch Schlaf und die Synthesis der Wachperioden etc."

D 1 (1932) "Allgemeines über Theorie der Erfahrung vom 'Realen'"

D 5 (1917 or 1918) "Zeit.—Dekungssynthesen (Übereinstimmigkeitssynthesen), passiv, und ihre Korrelate.—Zeitmodi und Urteilsmodi.—Glaube (belief). […]"

D 8 (1918—Bernau) "Individuation, das 'Tode-ti'"

D 14 (1931-1933) "Wachheit, Schlaf, Unbewusstes"

Acknowledgements

There are a number of persons and institutions I wish to thank for supporting, and inspiring, my research. The first are my teachers: Erazim Kohák, whose courses at Boston University when I was an undergraduate steered me in the direction of serious philosophical research in phenomenology; and Krzysztof Michalski, whose graduate seminars on Husserl and Heidegger were of decisive importance to my understanding of phenomenological philosophy. I also wish to extend my gratitude to Klaus Brinkmann and Daniel Dahlstrom, both at Boston University, who have been generous supporters of my work.

The main lines of this interpretation of the *Crisis* were developed in a graduate seminar I taught for the Political Science Department at Boston College in 1999, which was supported by a fellowship from the Olin Foundation. Further research was pursued in 2000 at the Husserl Archives in Cologne, Germany, which was supported by the Alexander von Humboldt Foundation. I wish in particular to thank Dieter Lohmar for his help in finding my way around Husserl's unpublished manuscripts, and Klaus Düsing for his sponsorship of my Humboldt fellowship. During the preparation of the final draft, I benefited greatly from the opportunity to teach a lecture course at the Graduate Faculty, New School University, on Husserl's *Crisis*.

But above all, I owe great deal to a long series of philosophical conversations that I have had with Nicolas DeWarren over the past fifteen years; there is very little here that did not find its first formulation, and inspiration, in those discussions.

Introduction

Science and Reflection

> What is the value of form in life, the life-creating, life-enhancing value of form? A gesture is nothing more than a movement which clearly expresses something unambiguous. Form is the only way of expressing the absolute in life; a gesture is the only thing which is perfect within itself, the only reality which is more than mere possibility. The gesture alone expresses life: but is it possible to express life?
>
> —Lukács, *Soul and Form*

Reflection as *Besinnung*

What is the purpose of Edmund Husserl's last work, the *Crisis of the European Sciences*, and what makes it stand apart from his other texts? On one level, its purpose is the same as all of Husserl's other published works since 1913: to provide an introduction to phenomenology. Nevertheless, the *Crisis* is unique among these works, for its approach, for the first time, is historical, though in a special sense: the sections devoted to the history of philosophy are not merely an historical prelude, but represent a reflection on phenomenology itself, on its status as philosophy. History serves the purpose in these pages of opening the way for a critical introduction to phenomenology, and its elaboration takes the place of a straightforward elucidation of concepts and theories. The insight is that there is something about phenomenology that can be understood best through a critical-historical reflection. Yet the status of a critical introduction, even one that makes use of historical reflections, is something the *Crisis* shares with other works of Husserl,[1] including the first volume of the *Ideas*. Nor does the cautious, preparatory character of the *Crisis* set it apart: one could argue that an introduction to phenomenology was for Husserl always at the same time an introduction of oneself into philosophy, which means actually taking those initial, uncertain steps necessary for genuine philosophical awakening, and not simply sketching a set of recommended methods for the edification of the reader.

Perhaps what really distinguishes the *Crisis* is that never before had the question of the spiritual situation of the age, of the state of European cultural life and its relation to philosophy, been so central to the argument. This is certainly true to a large extent, though such questions had been important for Husserl's self-understanding as a philosopher long before he sat down to write the *Crisis*.[2] One finds for sure a new sense of urgency, though by no means panic (something that should be noted in advance—the *Crisis* is not a work of despair, but one which, if anything, holds true to a course Husserl had been exploring for decades). This is so much the case that in order to understand fully what Husserl has achieved in the *Crisis,* the temptation should be avoided to allow considerations of the cultural and intellectual climate of Germany on the eve of the Second World War to overly dominate the interpretation of the text. Some caution should even be exercised when approaching Husserl's work from the vantage point of the history of philosophy, a point of departure that Husserl's line of thinking in the *Crisis* itself strongly suggests.[3] For in the end, I would argue, the *Crisis* is more than a reaction to the times in which it was written, and more than the argument that phenomenology has an important place in the history of philosophy. Its chief concern, and with that its center of gravity, is the apparently more modest attempt to critically introduce phenomenology; and it is in light of this task that the problems of addressing the "question of the times" and the historical importance of phenomenology are posed.

However, if one were to set aside these two possibilities as candidates for guiding an approach to the *Crisis*, then there arises a need for another concept, another point of departure that will allow the *Crisis* to be understood on its own terms. I would like to suggest that the best concept for this purpose is "*Besinnung*," a term often translated as "reflection."[4] In many passages, Husserl describes his project in the *Crisis* as a reflection, in particular a reflection on science; and it is through such a reflection that Husserl formulates a critical introduction to phenomenology. Yet that does not mean that this introduction, which I would propose should be understood as a preparation for thinking in a certain manner,[5] proceeds by first outlining a theory, or even a description of science. The point is not that, in order to study phenomenology, one must first articulate an idea (or ideal) of "science." Rather, the idea is that a critical introduction to phenomenology must first secure the path for a critical introduction to the idea of science, the result of which will be more than a working definition of the latter. This procedure is very much in continuity with Husserl's earlier work, from the *Logical Investigations* to the *Cartesian Meditations*: phenomenological philosophy finds its beginnings in a critique of science.

This means that the purpose of the *Crisis* is to be both a "reflection" and a "critique." Yet in what way do these two concepts fit together? Or does this question already suppose too much—namely, that reflection and critique are two different things, two separate activities that need to be coordinated? Perhaps they amount to the same thing. It could for example be argued that critique is a particular kind of reflection, or vice versa, that reflection is a particular kind of critique. For there is a sense in which critique is not simply a question that comes on the heels of reflection, but is a mode in which a question, namely the question of justification, can deepen reflection, by transforming it from within. But this can simply be taken as a sign that reflection was already a questioning to begin with, that the question of justification is really only a way of "turning back" to reflect on a question already in place, just making it more explicit and evident.

The first question is thus how to understand the relation between critique and reflection. For the purposes of exploring a possible point of departure, let us mean by "critique" that particular kind of reflection which, so to speak, gets to the "heart of the matter," revealing something in its most basic sense. Critique, then, is a reflection in which something is given as what it genuinely, essentially "is," where this givenness is understood from the beginning to be a determination of sense. This formulation is still vague, but it does express an intellectual gesture that is very characteristic of Husserl: an emphasis not only on the theme of sense or meaning, but above all on the place and importance of the experience of the insight in which we grasp something as "itself," in its "genuine sense."

What kind of reflection would get to the heart of science, to the point where we could call such a reflection a critique, in the sense of grasping the genuine sense of science? It cannot be a reflection in the narrow sense, the simple turning-towards something to consider it meditatively, as when one recalls what something looks like, or pauses to consider an idea or a point in a debate. To get to the heart, it must somehow be more searching; and to grasp the essence, it must be capable of assuming the form of an analysis that can distinguish the essential from the inessential. Yet this just comes around to the same ambiguity cited above: if reflection were to become a term for a kind of critical analysis in which science would be broken down into its constituent elements, thereby opening the way for the justification of each, then this would clearly be stepping outside of what can be reasonably designated by the word "reflection." Reflection is not yet analysis, even if analysis moves within a domain opened by reflection.

Still, must not critique, real critique, be analysis? The answer seems to be yes, that critique requires analytic instruments of distinction and identification in order to fulfill its function. Still, there seems to remain a way to think of the act of reflection itself as critique, and not merely an opening move employed by a critical project, or perhaps even an instrument of criticism. Is there not a sense in which reflection itself is already a project, already something that would in turn call for its own instruments?

In order to explore the question of reflection, it is helpful to recall that the German word *Besinnung* has a somewhat different set of uses than the English "reflection." For example, *Besinnung* can imply the effect or presence of wit, of being about one's wits. In a related use, it can simply mean "conscious," in the sense, to recall some English colloquialisms, of someone who is "with it," or "in the game." Likewise, the word *Besinnungslos* should not be translated by the phrase "without reflection," but instead with something like "out of it" or "witless." The phrase *zur Besinnung bringen* means to bring someone around, back to one's senses. Then again, the word can indeed be used to mean "reflection," but often in the somewhat modified sense of "to become aware of something (again)." This is more than just an etymological point, and by no means would I wish to make an argument from etymology alone: it is in fact how Husserl uses the term in the *Crisis* and in many other texts of the period, where he emphasizes the connection of *Besinnung* with becoming-conscious.[6]

To be sure, the connection with the theme of consciousness does not immediately clarify matters. It does not answer the question why Husserl would use the word *Besinnung* as if it were synonymous with critique, something he does even in texts before the *Crisis*.[7] For it would seem that, by "reflecting," one does less than what one would be doing were one to instead give a full analysis, or theory, of science. Self-consciously embarking on an analysis of science involves the employment of often rather sophisticated concepts and analytical procedures to evaluate what happens in scientific activity. But to "reflect" in this apparently still shallow sense of simply "becoming aware of something" would seem to be only the first, modest, and least advanced step which, even though it perhaps must be taken, nevertheless does not go very far. So why stay with such a heavy emphasis on reflection?

Let me begin again, and attempt to take the question in a somewhat different direction. Starting with a robust association of *Besinnung* with consciousness as the guiding clue, consider the following thesis: for Husserl, a *Besinnung*, a reflection, is needed in order to become conscious of science. To be sure, in this direction there are also questions: what does "to become conscious of science" entail? More: of what about the sciences are we not already "conscious"? Yet I would argue that these questions open more possibilities. Clearly, we are conscious of science, insofar as we are aware of its place in our lives as that which provides us with an explanation of things, and which is often even identified as the very project of human understanding itself. However, to say that there is a need to become conscious of science does not necessarily mean that we are not aware of what science does, or the role it plays in our lives. Rather, the suggestion is that the familiarity with science, even that of practicing scientists, is not the same as being fully aware of the meaning of science. One can have a good understanding of what the idea of science implies, and this understanding can in turn guide activity, but this familiarity need not take the form of an explicit grasp, or consciousness, of everything operative in the sense of science. It is thus "non-genuine" in this sense, namely that part of the meaning of science remains hidden; its full sense is unfamiliar, a latent possibility of what is familiar—all without compromising understanding. Heidegger makes a similar point at the beginning of *Being and Time*: the absence of explicit knowledge of the meaning of being, the foundation for the science of ontology, does not exclude the possibility of the development of ontologies from within the horizon of what is partially understood with respect to being. On the contrary, the forgetting of being, or the obscurity of its sense, may very well be the condition for the emergence of scientific ontology as such.[8]

It is precisely the "full meaning" or the "full consciousness" of science that is the object of reflection in the *Crisis*.[9] Such a reflection is necessary because not everything that is operative in the understanding of science is explicit, articulated. Yet the issue here is not limited to being able to say, with the legitimacy of knowledge and understanding, "this is a science," if by that is meant being in the possession of a clear definition. The full extent of the meaning of science surpasses its definition, however accurate, as well as the set of conventions that must first be mastered in order to be able to identify a field of study as a science. Neither of these expressions of meaning (definition, convention) exhaust the essence of the understanding of science; above all, a sense for the significance that science carries for life is left open. It is this importance of science—of its meaningfulness in the sense of weight, value, or what Husserl calls "evidence"—that is uniquely manifest in reflection.

The claim of science

In a preliminary fashion, one of the key insights of the *Crisis* can be formulated as a double thesis with respect to the meaning of science: (1)

First, the value of science, not always expressed, but nevertheless operative, has to do with an implicit *claim about meaning in general.* Namely, science represents a claim that, for something to be meaningful, it must take a particular form—it must be *rational.* Leaving aside for the time being what is meant by "rational," it is instructive to emphasize that an essential aspect of science is that it represents a kind of attitude, or positioning with respect to the theme of meaning as such; this is its genuinely "logical" character, as Husserl argues in *Formal and Transcendental Logic.*

Second, (2) a full grasp of the "meaning of science" includes a comprehension not only of the directly asserted content of the claim of science—what is meant by "rational"—but also what it means to make, but also to value such claims about what is to count as meaningful and non-meaningful at all. In light of this second thesis, the question of the importance of science can be brought into focus, for here the theme is not only what science claims, but also what is at stake, or that to which the claim is being addressed as a response. These two theses are of course intrinsically related, and they are not unique to Husserl; one could even say that they necessarily provide the subtext to any discussion of the meaning of science, or what science is "ultimately about." To talk about science at all brings both into play: there is the meaning of what is being said, and what it means to say it. Yet what is unique about Husserl's reflection on science in the *Crisis*, I would argue, is the sophistication and insight with which the question of what it means to assert the claims of reason is explored. Husserl, in his quiet way, not only defends reason in these pages, but also has the honesty and discipline to deeply question the potential significance or insignificance of such a defense.

If part of what is at stake in a claim about meaning in general is the way in which the very possibility of making a claim at all is to be understood, then the question is in this sense "foundational." However, when speaking of foundationalism in Husserl's philosophy, it should be kept in mind that the foundational by no means excludes questions, or even doubts of legitimacy. Certainty need not be unassailable in order to be foundational. It is clear that Husserl is convinced that the claims of reason are justified, even absolutely so; but his convictions did not blind him to the consequences of the fact such claims are precisely *claims*, and that claims are always exposed to the call for further justification. Husserl was very sensitive to this aspect of reason, and developed a powerful descriptive vocabulary in order to give it philosophical expression. The result is a kind of "foundationalism" that has a demand for renewal, or re-founding built into its very structure.

Another result is that Husserl's kind of "foundationalism" puts one in a position to ask, in a rigorous manner, the following question, which will guide much of the presentation of Husserl's last work below: does the meaning of the claim of science—in the second sense, the meaning of making and valuing claims about meaning—itself instituted, and with that operative only in the form of a claim? Is science ultimately founded in a claim about what it means to make a claim (about meaning)? This question is not simply premised on the exercise of pushing towards a "final grounding," but on the real possibility that the ultimate claim of science, what it cannot part with and still represent a possibility of life, is not any given articulation of the "idea of reason," but a claim that the *question of reason* as such demands our attention and efforts to understand.

It can be said, I would argue, that one of Husserl's important contributions to philosophy is a reflection on the horizon in which the claim of reason is constituted. Different aspects of this contribution can be followed not only in the *Crisis* itself, but also in a number of texts from the late 1920's and the 1930's that have recently been published as an *Ergänzungsband* to Husserliana VI.[10] Husserl's engagement with the meaning of science in these texts is not limited to an evaluation of a particular, historically articulated idea of reason. His project is to reflect on the varied emergence and constitution of the act of meaning formative for the experience of scientific thinking, in order to discern the teleological unity of its historical variations—thus to reflect on what we aim at, when we speak "scientifically," where this "we" designates a trans-generational community of scientists and philosophers. The value of the question itself, in other words, is intrinsically linked to the historical emergence of a particular kind of life, which means a particular mode of being conscious of the world. And if, as I would like to suggest, it makes sense to ask whether the value of the question of reason is experienced in the mode of a *claim* to value, then perhaps a consequence of Husserl's analyses could be the thematization of the way that, through its historical genesis, a certain mode of life makes its claim on us, addressing us with respect to the task of its appropriation.

Critique and *Besinnung*

Husserl understands the task of appropriation as the task of critique. Yet to rely on the conception of "reflection" in order to characterize Husserl's thinking in these texts may give rise to a certain unease; one may have the sense that all reflection could do is set up the theme, or at most a line of questioning, and that it will never really arrive at the level of a true "critique," or even a real understanding of what is at stake in science. For in order to understand, even to understand a claim, are we not at some point compelled to engage other modes of thinking and analysis than reflection? For example, to truly engage science, must one not, at some point, deny its validity, or at least challenge it, precisely in order to comprehend why in the end it is a claim too important to be denied? Once the theme has been introduced of a claim made by "scientific consciousness" on contemporary humanity, does not Husserl thereby move beyond reflection? To remain within reflection would seem to stall philosophical engagement at the level of indifference; but can such indifference at the same time be compatible with critique, in particular with respect to the question of what is important and unimportant?

In fact, Husserl never sharply distinguishes the activity of philosophical *Besinnung* from the choices that must be made in the wake of reflection. One reason for this is that reflection in the sense of *Besinnung* is not premised on an indifference to what is being reflection upon; quite the contrary. And the non-indifference of *Besinnung* keeps open the possibility of its functioning in the mode of critique, at least in the case of science. Another way to put this is that for Husserl, to become aware of a sense or meaning has immediate consequences on the shape of belief; a separate activity of drawing consequences from a completed reflection in order then to shape beliefs is not necessary. This is because awareness is always already a mode of engagement, one that takes aim at some goal; this is as true of the awareness of clarity as it is of the awareness of purpose. In turn, this implies that the consequences of an awareness are not always external, but are sometimes manifest in the form engagement takes once it has developed into something substantial, or complete.

Thus if the essence of science is a claim about what is meaningful, then a completed reflection, as a particular engagement with this claim, can be said to have *become* critique, as an inner consequence of a reflection on sense shaping belief. Critique, then, would mean the shape reflection gradually takes once it has been set into motion, once it has been exercised; this fits well with the experience of reflection, where an object only gradually comes into focus. Critique thus understood is the maturation of reflection, the realization of the coming-into-focus of an object that was initially the vague, indeterminate goal. More, in the case of the meaning of science, reflection is in a sense *driven* towards critique: for the claim of science can only be clarified if reflection finds its way into belief, otherwise the claim, as a claim, would remain vague and indeterminate. To engage the meaning of science can only be to engage the question of its legitimacy—for what other form could the engagement of a claim take, if not in the form of the question of its legitimacy, of the possibility of conviction.

One might stop here, and say: has this line of thinking not led to a contradiction? Before it was argued, that reflection does not aim at evaluating the claim of science; here it is said that reflection necessarily evaluates the claim of science. Yet it is not the case that an engagement with the question of legitimacy must take exclusively the form of an evaluation, which would have as its outcome a yes or a no; it can also take the form of the question of responsibility, where what is at stake is whether or not I will give my voice to what is being claimed. Responsibility is one of the key themes of Husserl's texts in this period, where it is given more weight than even that of critique.[11]

To be sure, the question whether or not to take responsibility for something, to speak on behalf of it, does not exclude evaluation. It is just that there is more involved: to give one's voice to something, to stand for it and by it, is very different from affirming a simple yes or no, confirmation or denial. A proposition can be confirmed without any role played by my responsibility; it can confirm itself, in the sense of "turning out" to be true; I can let it speak for itself, or speak of it from a position of anonymity. I can, that is, reduce its "claim" to something that does not directly touch on me, as if its being asserted is not because it has made a claim *on me*. However, if I become responsible for it, then it becomes my own, in an important sense a part of my consciousness of self.[12] Thus to reflect on the possibility of making the claim myself, in my own voice, not only brings the truth of a proposition into question, but it also brings my self into question as well—for the question here takes the form: what would it mean, to be the one who would make such a claim? The idea, basic not only to Husserl's philosophy but to many other phenomenological philosophies, including Heidegger's, is that there is a sense in which the engagement with meaning entails an engagement with the question of one's own possibilities. And more: engagement of this type cleaves close to the sphere of self-consciousness, and cannot survive a full translation into the indifferent, third person modality of evaluations. Thus there is no contradiction here; one can still hold that reflection does not evaluate the claim of science, but that it does engage it as a possibility of responsibility, a possibility that is genuinely manifest only within the sphere of self-consciousness.

To reflect on a claim in light of the possibility of making it one's own, as one's own possibility, means that it is not meaningful to consider the claim as foreign, or coming from the "outside." A basic condition for a successful, fully objective evaluation is thereby excluded. Instead, to engage the possibility of speaking on behalf of this claim, it is necessary to begin a process of attempting to recognize oneself in the claim; but for that to occur, the claim must be given shape, an articulation of such a kind such that it can take on the sense, in reflection, of something that *comes from* oneself. I have to put it into my own words, trying to understand it as if it came from myself. If this attempt were not made, then the entire question of appropriation and responsibility would simply be arbitrary; at worst, one would be a collector of claims, imitating what is said by others, or at best, a spokesperson for someone else.[13] Reflection would then indeed fail to be a serious engagement with the meaning of science; it would have no connotation of a "consciousness" or a having "come to something" on the part of an awareness. But that must also mean that the movement of appropriation, within reflection, is in fact a kind of establishment: for in reflecting on the claim as a possible form of my own life, I in effect seek to re-create the claim in my own person, precisely in order to see in it something that would recognizably belong to my life—or *could* belong, for attempts to articulate, to recognize, and to see can always fail. Such failure does not need to be a clear case of misunderstanding: there are some truths, perfectly coherent and understandable, which are nevertheless not "defensible," in the sense that they are "not worth talking about," that they may simply have no place in the repertoire of one's convictions and responsibilities.

It is also clear that in reflective engagement of this kind, the meaning of science is not left unscathed. It, too, suffers consequences thanks to the role of this articulation for oneself in reflection: meaning becomes re-established, which implies being given a new shape. This is a central thesis of Husserl's, and it is not unique to the *Crisis* texts.[14] Yet the thesis is not that, in reflection/critique, we create meanings; but nor is it the case that we are simply testing the validity of opinions about which we happen not to be certain. Instead, for Husserl reflection is the medium through which the meaning of science develops as such, in an important sense *for the first time*, and for two reasons. First, (1) it is only in reflection that the meaning of science, as a claim about meaning in general, makes its appearance as such, and is thus intended as something with which we are engaged, for its own sake. Prior to reflection, the claim of science, though operative, is not necessarily visible; it can remain in a twilight of naiveté, quasi-present, but not expressed. But meaning cannot be what it is in such a latent mode; in a way, science has no meaning before reflection, it is not even science until its meaning is finally expressed. So for example, as Husserl argues at the very beginning of *Formal and Transcendental Logic*, the "science" that existed before the critical reflections of Plato is not what we, who think *after* Plato, call "science." The critical formation of sense in reflection is thus an original institution of sense, an origin, that is not immune to repeatability. Second, (2) it is only in reflection that the possibility of responsibility can be formulated with respect to the claim of science, where an explicit awareness is developed of what is at stake in the question of its legitimacy. It is thus in reflection that science acquires its very sense, for here it finds its form; and for the Husserl of the *Crisis*, this form is that of self-responsibility.

The "grip" of meaning

Yet for Husserl such considerations of the meaning of reflection cannot remain on the level of form (that in reflection science takes a certain form it does not take in "naïve" life); that would not get to the heart of the matter. Alternatively, the issue of responsibility is not only one of sense or meaning, but of truth. Thus what is lacking, so to speak, is the soul of science, what animates it—and this is not something limited to the sphere of reflection, which means that one must, as Husserl himself does in principle, orient the discussion not only around the experience of reflection but of the world itself, of life. Our reflections should always be guided by our common experience, for there is always another side to the experience of meaning than that of form alone. In Husserl's language, any awareness of meaning, whether established through reflection or not, is always something *motivated*, it has its "evidence." Awareness is motivated not (only) by desire, or a practical purpose that would give an impetus to inquiry, but rather in the sense that the awareness itself, once achieved, results in a certain "hold" or "grip" on us, a passive being-moved into a particular position with respect to what we experience as meaningful.[15]

The motivated character of meaning is multifaceted. As motivated, meaning becomes an abiding acquisition, with a sense of permanency; as such, it becomes the basis for traditions, where meanings have a feel about them of a natural concreteness. Traditions last, insofar as their presence is not wholly dependent upon the free acceptance or rejection of a given thesis or set of theses; their hold is to some extent already accomplished, already in place before one decides how they stand with respect to it, if a decision is made at all. Another facet of the motivated character of meaning has to do with possibilities. Only as motivated can a meaning be a real possibility, as opposed to a "merely logical" possibility. The difference here is not a question of form. A possibility is a possibility only in contrast with an actuality, which effects the contrast by filling the place once inhabited by the possible. One might be tempted to refer to a notion of an "empirical possibility" to capture the sense in which a possibility inhabits such a place in the absence of an actuality. However, the question in this context is not really one of how to assign possibilities and actualities to places. The sense of the concreteness of the empirical will always remain elusive until some manner is found in which to express the way that the possible (or for that matter the actual) holds us fast to the particular context to which it in turn gives shape.

For example, in any situation in which we may find ourselves, be it the most mundane situation of ordinary life, part of our awareness is a sense of being confined within the curve of a certain inescapability. Action, life itself would be extinguished if certain possibilities that define, here and now, what can happen "next" were not addressed, whether by an agency or the unfolding of events generally. Thus a possibility is concrete not only because it holds a place open for an actualization, for something to occur, but also because it holds one fast to the "moment" as a necessary element of the situation as such, a kind of limiting focal point of our concerns. Something must happen, something must be satisfied in order to exit any particular situation; thus to be aware of the situation at all, to be "there," is in this sense to be in its grip, or situated. Moreover, this means that to be unaware of something is, at least on one level, to be out of its grip, free from its influence or motivation.[16] Motivation understood in this way represents just as much of a constraint as it does a release, a hemming in as much as a drive onwards; to fix the parameters of a motivation means to fix both of these essential aspects.

How this bears on the world of ideas can be illustrated with the story by Jorge Luis Borges, "Funes, the Memorious." It is about a man named Funes, for whom there were no limits either to his attention or to his memory. He never missed a detail or aspect of anything that he saw, heard, or felt; and his memory, in turn, never faded, never let go of the unlimited richness of the moment captured by his attention. Consequently, Funes becomes a helpless victim of his own perceptions that, so completely filling his consciousness with all the minute details of the world, made it impossible for him to "think," in the sense of finding a place in his awareness for the articulation of generalities and concepts. Funes never formed what Locke called "general ideas," because there was never a situation for him in which such ideas could take hold. Their very possibility made him uncomfortable: "He was disturbed by the fact that a dog at three-fourteen (seen in profile) should have the same name as the dog at three-fifteen (seen from the front)."[17] Present to Funes were the memories of each and every single particular instance of all the perceptions of all the dogs that he had ever seen, forever available to him in all of their details; nothing forgotten, nothing let go of, there was simply no satisfying way to classify particulars under general concepts, no motivation to see the particular through the general, or to consider the general in place of the particular. The superhuman attentiveness of Funes opened him to the sea of details in the present and the flood of details from his past; but the hold that each detail had on him constricted the very horizon of his possibilities; only details had sense, only that which took the form of the concrete given could he "see" in a genuine sense. This is what Husserl means by motivation: meaning is motivated, because meaning is an instance of the grip of the world, structured in accordance with the possibles of a given situation, which in turn take the form of a limiting-opening.

Therefore, to be aware, or to be conscious, is not simply a matter of having a grip on something, but also of being "in" the grip of something. Bringing a meaning to consciousness—in the case of reflection in the *Crisis*, the meaning of science—is not a simple matter like telling a story (not that telling stories is always simple), or taking up this or that topic for consideration and study. Bringing a meaning to consciousness means, in part, to be placed in a possible situation in which we bring ourselves under the influence of meaning; it is to be limited, and with that captured, in order to be opened to a givenness of a particular kind and species.

Take again the kinship between reflection and critique, which I will try to make more precise. A reflection on science, in the sense of *Besinnung*, is critique in the sense that it aims for what could be called the *fulfillment* of the idea of science. In reflection, the idea of science is fulfilled in its articulation, where it is given a new shape, in a context in which it becomes something *actual* for those who discover it as a possibility of self-responsibility. There is here a more familiar notion of critique on the horizon: if in reflection we become conscious not simply of what science is, but what it means, what grip it has on us (or better: *can have*, as the result of critique), then presumably we are also in a position to become clear about the limits inherent to this motivation, or what remains impossible for science even after it has been affirmed in its essential possibility. For like Funes, what the meaning of science excludes, in order to open for us a particular way of seeing and experiencing ourselves and the things around us, may still somehow be present to us, and even perhaps represent a source of unease, as did the possibility of general ideas for Funes.

To be sure, none of this means that one cannot "merely think" about the meaning of science, or about any meaning whatsoever. Like with anything else, one can of course "merely" reflect, simply having "in mind" something that can then be "merely described." However, that does not represent, for Husserl, the most original relation to meaning. Meaning for Husserl is something that, in its primary lived sense, takes the form of that which is significant; there is a *weight* to meaning, a movement that always makes it more than just an "object," if by that we mean something to which one can essentially remain indifferent. When meaning is fixed as an object, something is neutralized which, in its original presence, has always already made a difference. Once neutralized, what are seen from the new point of view of indifference are the contours of a difference that the event of meaning had originally accomplished, or the impact that it had had in defining the limits and structure of an openness. This difference made by the event of meaning is what Husserl calls evidence (*Evidenz*), and its elucidation was ultimately understood by him as a genetic-phenomenological problem involving the central concepts of *life* and *world.*

Life and world

The problem of evidence is decisive with respect to the question of the origin of meaning in Husserl's later writings. At this point, however, it should be noted that these metaphors of "grip" and "hold" must also be understood as an indication of something that is only anticipated on the level of reflection, which amounts to an important limitation with respect to the use of the idea of reflection to guide an approach to Husserl's text. The issue is the manner in which abstraction plays a central, but ambiguous role in reflection. For when one reflects, the activity takes place on an abstract level; this is something understood from the beginning, and it is in fact necessary that such an understanding be in place in order to be able to begin to reflect at all. The consequence is that, if the meaning of science finds its fulfillment within the gesture of a reflection, as was suggested above, then this turns out to be equivalent to saying that the meaningfulness of science is possible—in a fulfilled, or "genuine" sense—only given a turn to the abstract. Moreover, it is precisely with this awareness of abstraction that a perfectly natural anticipation grows, namely, that the significance of what is discovered in the abstract can "spill over" into our experience of science in the concrete, imbuing it with a newfound sense of value. One anticipates that at some point this meaning of science, once it is present in an explicit awareness, will also have significance in the world of the interests and purposes of life in the concrete. Such anticipation is not unambiguous, however; and its justification, which amounts to the argument that the abstractions of thought ultimately have a place and function in the world of life, is an important feature of Husserl's thought. To "understand" a concept or idea for Husserl is always to think in such a manner that one can see in what way it is ultimately at home in the world of "life"—and in such a manner whereby one ultimately comes to the recognition that life is the only possible context in which meaning can be nition that life is the only possible context in which meaning can be "significant" at all.

As it stands, this thesis is fraught with ambiguity. "Life," but in what sense? My individual, concrete personal existence, or life as a more general principle of consciousness? More, in what sense does life have a world, one that lends itself to being, at least in part, defined by the claim of science, accepting that this claim is something about which we can first be clear only on the level of abstraction? How the theme of life and its world can be a legitimate theme of a *Besinnung* on the meaning of science is the chief task of any interpretation of Husserl's *Crisis*. Right away, it should be stressed that Husserl's conception of the "world of life" will not be that of a set of practices which, in order to ensure the achievement of a given set of ends, requires the employment of the faculty of reason; nor will the reference to the world of life be a device for calculating the "weight" or "value" of a given meaning (even abstract meaning) in terms of its "use," whether material or spiritual. Instead, the shift from a critique of the meaning of science to an understanding of the rootedness of this critique in life will involve at first the rather more modest claim of philosophical significance. Above all, insofar as it is at every turn anticipated by reflection (also something to be shown), the theme of life is not imported from the outside, but is integral to each phase of reflection. Thus, when beginning to read the *Crisis*, I would propose that one understand by the concept of "life" that which is naturally anticipated by abstract reflection, namely the promise of being able to, at some point, establish its own philosophical significance. Or, in other words, what is anticipated is the possibility of experiencing the genuine hold or grip of the meaning of science, where "life" is a term for the lived experience of being moved along the tensions opened by the limiting-opening of a concrete, established meaning.

Nevertheless, this more modest thesis is in fact more radical than it looks. For in the case of science, according to the double thesis about the nature of the claim of science sketched above, the anticipation of its significance involves the possible weight of a claim with respect to what is meaningful *in general*. Thus the possible fulfillment of the expectation is not limited to a success in having found a place in the world of life for the meaning of science, but involves the power of a claim which would project the outlines of a possible relation to life as such, or to the kinds of meaning that could structure life itself in a *universal sense*. What shows itself here is an intrinsic, though often invisible movement within any reflection: the projection of the possibilities of meaning, thus of a future shape of things.[18]

There is another side to this anticipation that is important to emphasize. For when reflecting philosophically on science, however sharp the turn to the abstract may be, it is not the case that one takes up an empty or open possibility that has not yet been articulated or fulfilled in any sense. Science, after all, already exists; it is an actuality that has already run through the possible, giving it shape and expression. Husserl does not make the claim that he is about to articulate the idea of science for the first time; its very existence has already been founded in and as its first, most original expression. In reflecting, one does not "discover" science, not even the possibility of the forms that the meaning of science can assume. Science is a unity of sense that precedes reflection; it is already in place, which means that what is anticipated in a reflection on science is also something that may already to a large extent be in place, even if its precise location and layout is not yet clear—not everything in the future is new, not every new form of expression is unanticipated. In general, however much it may be possible that, in reflection, science will indeed take on another shape, another form as a possibility of our responsibility, as a whole science is something that has always already found its place in the world of life.

Not only that, but science itself is already an accomplished *reflection* on its own meaning or sense; it is already an example of the gesture in which it will be reconstituted when reflection is taken up in the mode of *Besinnung*. What is more, in the wake of this reflection, the experience of the world has already been (re)formed, instituted by science as a human activity, which means that it is already something that has been lived. That arc through the abstract back to the world of life has already been traced, and in a way, Husserl is simply trying to find a way back, in order to re-experience the originary act of thinking to which science owes its emergence. The result is that, however "abstract" the reflection in the *Crisis*, the point is not to approach science as a mere logical possibility, at least not if one is to see whether it could become something of significance, not for the first time but in a new and unfamiliar, if not completely unexpected way. In other words, through *Besinnung* we approach science as an idea of what it is to fashion an idea of things, where this "idea of things" has already occurred, already taken place as a concrete possibility of life; this means that the question of the meaning of science is not an empty exercise of form, it is already a gesture of a philosophical life, and thus is already a concrete matter in the context of motivations. In short, one must be a philosopher in order to reflect on what it is to be a philosopher; it is, in a sense, an activity that is at least partially invisible from the outside.

Along with this distinguishing characteristic of Husserl's reflection on the meaning of science, there is an important point to be made of a more general nature. For Husserl, meaning cannot be reduced to propositional content, to what "words mean" or how they are used; meaning is what it is only with reference to the movement of a life for whom it is "carried out," "relevant." Meaning is thus dependent on its articulation, not by an act of speech but by experience; it must be taken up into a movement of lived actualization. And lived experience, in turn, depends on its meaningfulness; a lived experience that articulates nothing, that forms no content, leaving nothing behind but the invisible trace of its passage, falls short of an "experience" as such. Thus the argument cannot be that meaning can be understood only if we place it within the world of life, because life itself in turn finds its place, the outlines of its very contextualization, precisely in the meanings that it articulates; there is a complete interdependence here in which there is no room for indifference on either side. For Husserl, the world of sense, of lived meaning, is not a universe parallel to that of a "concrete" life of events and objects; it is itself the medium in which life pursues itself, in which it finds the space within which to realize the unity of its "reality."

If so, then to ask what it means to say that something is "meaningful," the reference to what in life is considered to be important would not in itself be adequate; it would be misleading to say that "human life," more real than meaning and more actual than the reflecting subject, provides a standard for judging what is important and not important. The most that can be found in life is, as was suggested above, a set of claims. In a sense, when one considers meaning as it appears within the horizon of life, it is this very appearance that *is* its importance; for there is a sense in which visibility is intrinsic to validity. One of the consequences of Husserl's phenomenology is the demonstration of visibility as a primordial form of validity, one that is essential in order for meaning to become established. This is worth stressing in advance, since however important the notion of "life-world" may be in the *Crisis*, it is not a substitution for the central concept of *phenomenon*. All validity and value refers back to the phenomenality of the given, and the constitutive accomplishments of consciousness generating the fold of this phenomenality.

Yet this does not mean that Husserl's conception of life in the *Crisis* is merely that of a kind of showcase for subjective accomplishment. Life is, instead, understood as the horizon in which an interest is developed, a directedness towards the given in its phenomenality; it is in and as life that an interest in what can be made visible takes root, an interest not only in visibility but also in the validity that comes with the possibility of seeing. Life, in other words, situates all the accomplishments of phenomenality. Seeing is in a way like speaking; for one speaks not just for the sake of speaking (at least not most of the time), but for the sake of saying something *with meaning*, thus for the sake of a speech understood as having a weight or impact in the situation in which something or other is addressed—or, to invoke a Nietzschean turn of phrase, for the sake of a certain *value* that orders the world. The articulation of meaning, whether with the eye or the voice, is for both Husserl and Nietzsche a fundamental *valuation*, one that can be understood only in terms of an *interest* in a meaning and an order, or better: a *claim* voiced on behalf of their possibility.[19] The seeing that comes with all phenomenality is thus not a simple transparent gaze, but carries with it all of the opacity of interest. This means that the claim that science represents does not somehow lie outside of life, accessible only by way of an abstract reflection, but is an articulation of the world by a certain kind of life, a particular mode of interest in things. To become conscious of this claim is to in turn become conscious of this life; thus the maturation of reflection as *Besinnung* implies the thematization not only of science as meaning, but of the world of interests in which this meaning ultimately has its place.

Thanks to this conception of life, Husserl could be seen as one of the last philosophers of the Enlightenment.[20] If "science"—or, to use more explicitly the language of the Enlightenment, if "reason"—can be shown to have an intrinsic value for life, that it is formative of its interests, then this would be equivalent to showing that reason takes the natural form of a *demand*, a necessity. What else is a claim with value, if not a demand? If this can indeed be shown, then the "claims of reason" would not be limited to the correctness of propositions or theories; reason in the form of a claim, prior to the world (for it is a claim about what the world is to mean), would itself be shown to be a demand on and for life. In short: to become conscious of the meaning of science would mean to put oneself in the position in which it first becomes possible *both* to make the argument that reason is necessary for a "good life," or a "genuine life," *and* to speak on behalf of the demand that such a life be realized. If reflection in the sense of *Besinnung* is the means by which the claims of reason are to be articulated, given new form, and advanced in the language of a project of establishing an order for life, then Husserl's philosophy is, at least potentially, of deep social and political significance. We could perhaps say that Husserl's reflection on science in the *Crisis* constitutes an attempt to rekindle that conviction and, above all, *self-assuredness* of life that had once found so forceful an expression in the cry "sapere aude!"

Husserl was convinced that such a philosophical gesture was still possible; he was also convinced that phenomenology points to the possibility of a renaissance of a life in truth for which such a gesture gives form. On the other hand, Husserl was also sensitive to the almost overwhelming odds against such a renaissance. The prospects, after all, do not look promising, given the characteristics of that kind of life formed by the claim of science. For this path of understanding, or of the experience of the world, requires an attitude for which sense and meaning must be perpetually prepared, established, constructed, proved, legitimated—in short, it is a life for which meaning is everywhere a question, a problem. In the end, self-assuredness here really has no place; it is not possible to be content merely to cultivate one's garden. That the problematic character of all meaning for scientific, rational life infects the very meaning of science itself is something to which the concept of *Besinnung* itself testifies: one begins fully aware that whatever will result from reflection, of careful investigation and clarification of sense, is from the beginning uncertain, and that it is necessary to be attentive to the problem of the very "becoming conscious" of what it is that is being made a theme. The fact that thinking must proceed abstractly also implies that it must proceed timidly: simple, obvious questions must be designed in order to avoid all presuppositions, as if the subject were something strange and alien, or as if reason itself were an instrument that had never before been used; and one must always be on guard not to miss a chance of catching a glimpse of some rare animal—even if, in the end, the hunted is one's own life, that which is "closest." There are countless pages of Husserl's working manuscripts that are characterized precisely by this careful watchfulness of the hunter: the same themes, descriptions, theses are repeated over and over again, tracing and re-tracing the same territory in hopes of catching an evasive thought. What sort of life is this, what sort of philosophy, which can only proceed in such a complicated, even tedious manner? And in turn: what sort of world is this, which is meaningful only as a problem, a question that must be approached so cautiously? What is the relation between these two questions? "What is the world?—What are we?—Why this double question?"[21]

As was stressed above, science is already something "in place," which also suggests that it already has meaning for life. For Husserl, philosophical reflection is not meant to give science meaning, but to give this meaning a new form. However, as we will be seen in detail in the presentation of the argument of the *Crisis* below, the most basic "hold" that the claim of science has on us is primarily that of a problem. Science has a hold on us the way that a problem has a hold on us—it both reveals yet keeps hidden, opening up a path of understanding from a point of departure in a failure to understand. This presupposes a particular perspective in order to have an effect at all. Science would not be possible within the closed world of a natural naiveté in which human understanding of the world is never approached as a problem, or experienced as a failure that, nevertheless, promises a clue. For naïve life, there may indeed be problems that call for sophisticated techniques for their solution, but the motivation to abstract the activity of thinking from all of its accomplishments, all of its familiarities, and consider it as a uniquely formed object in an attitude where nothing is taken for granted, all in order to see as clearly as possible the very possibility of knowledge itself—this implies being in the grip of a very different type of meaning than anything that could be found in a world and a life that were not in question.

This line of reflection already begins to show that, in spite of its proximity to the spirit of the Enlightenment, the purpose of Husserl's philosophy is not simply to make a case for "science" or "reason"—even if, in the wake of Husserl's legacy, such a case may very well be made. Instead, the explicit task, I would propose, is again that of reflection. The task is to become clear about something that already has meaning, thus its place. In the case of science, what has meaning, or can again have meaning, addresses us not as those who would simply accept it, "live" it in a naive fashion, but which, in order to fully be what it is, requires that we engage in a reflection that abstracts us from any possibility of simply "affirming" the pull of science as a "natural" part of our life. The meaning of science, of its claim, can become real, "in effect," only as the result of a reflection that is not incidental, or something engaged in only for edification and education. However much it may be the case that science is already in place for life, it is not there just to be learned; it must be rediscovered in light of its purpose, precisely as something intimately our own. Responsibility for Husserl does not result from a demonstrated autonomy, but is formative of the very possibility of reason.

I would also like to suggest that this point is of no small importance for understanding Husserl's *Crisis*. The relation of meaning to life, which Husserl articulates with remarkable consistency throughout his long philosophical career, by no means amounts to an argument that, where conceptual thinking is hopelessly obscure and incompetent, one can find in "everyday life," "as we experience it," all the clarity one could want. On the contrary: the understanding that we are, that is accomplished in our living, does not enjoy the simple givenness of a primordial clarity to which we can appeal whenever we run into conceptual difficulties, as if there were a store of common wisdom waiting to be mined by perspicacious philosophers. The very move to reflection itself, towards the development of an inner life that pulls itself back from the naïve acceptance of the world, is premised on a skeptical attitude that has always already infinitely extended what is meaningful beyond anything that could be immediately, directly, and simply "given." One already senses, even before engaging in philosophical reflection, that the subtleties of sense hint at a depth that, if it is to be opened at all, requires that one learns *how to see*; not simply to recognize and agree about truths that everyone, on some level, senses to be true, but to become aware of something that is not obvious but hides behind the obvious, and which can so easily be missed altogether. Reflection is the extension into criticism of this struggle to see; and what reflection reveals, and is necessary in order to reveal, is a horizon of significance that is open only once we have learned to live in it. Thus when Husserl speaks of a rebirth of philosophy as a "transcendental experience," or a "transcendental awareness," he is very serious: the new form of theoretical reflection inaugurated by phenomenology opens a new dimension for lived experience itself.

The theme of history

For Husserl, the emergence of the claim of science takes the form of the generation of a new mode of historicity (*Historizität, Geschichtlichkeit*). Insofar as reflection is the attempt to uncover the possibility of the grip of the meaning of science taking a new form, it can do so only if it is able to navigate the sense of history which determines the genesis of the claim of science.[22] The problem of history is central to the *Crisis*, and its elucidation is one of the most important tasks of any interpretation of the text. My strategy in what follows will be to outline the way that the theme of historicity in Husserl's *Crisis* is employed in order to clarify the kind of "hold" that the meaning of science has on life. Namely, the hold of science will be comparable to the hold of history itself, and in two ways:

(1) First, the meaning of science has a hold on us the way that the past has a hold on us. The past is "there," a part of the world; but it is not a simple given, a surface without depth. More, the past is not fully present, but represents a certain half-open, partially accessible space in experience. This space is there in the way that the unconscious is there, or a memory that has not yet been set into motion. All that is needed is a thread to bring it out, to begin to reveal what is latent—like Proust's little madeleine that sets off a search for "things past." So too with science: there is always a certain depth already present in its functioning, an implicit content that can be awakened given the right thread. For example, to engage in scientific activities, or even simply to refer to science as a body of knowledge, is to make use of scientifically achieved results, which in turn refer to an historical space that would require an explicit project of historical reconstruction to uncover several categories of elements that are normally only implicit: conceptual presuppositions, historical narratives that explain the pe-

culiarities of certain formulations, or the general sense of the nature of knowledge as a gradual cultural acquisition.[23]

This is true of any cultural phenomenon, perhaps the best example being that of language. Language clearly has a hold on us, we depend on it to articulate our very thoughts, even to ourselves; but it is something that, despite the fact that it is present for us in its constant use, is always more than what any human being, past or future, can claim to be his or her *own*. What I say when I speak, thus what counts as my words—these are my "own," but only because I have made them my own for the purpose of understanding and of being understood. The English language itself is not my own, nor is it something I have made my own; it is nothing that I have created or even intend to create in any sense. As such, being the purpose of no one, language can be created only in the historical space of a people, or what Husserl calls a "nation," which for him has the characteristic of a kind of "person."[24] When one learns to seek out what can be found in this space of experience in which the purpose of no one particular subject is pursued, one begins to catch a glimpse of a horizon, not of life, but of that within which life itself moves. Moreover, it is only because of this historical depth, only because of the presence of the horizon of the past in us, that we can have something like a "language" at all, giving our thoughts the expressions they require in order to be thought.

The same is the case with science: it can have its hold on us only given a sensitivity to, and thus situatedness in, this historical dimension of our life. Science, as a thinking that has found its home in language, is a meaning that would not be possible in a life for which only the perfect circle of the immediate present were the only visibility and validity. Thus, a reflection on the meaning of science for Husserl must be an *historical reflection*, for it is only for a consciousness with an historical depth that science "means" anything at all.

(2) Second, science also has a hold on us the way that the *future* has a hold on us. Again, we can point to language for comparison: what language gives, thus what can be received because of the historical character of life, is not simply a set of words, but a manifold of expressive possibilities. Meaning orders a field of possibilities, whether for expression or action; to relate meaning to life is in part to understand meaning as the projection of possibility and life as the fabric of the possible, which turns and twists according to the manifold projection of sense. A tacit understanding of this is contained in a number of everyday expressions: When something significant happens in someone's life, we ask: but what does that "mean"—for the relationship, for the job, for the future? Events have meaning "for" various ideas one may hold regarding the world, about oneself; to have

meaning is, again, to have consequences. Events in the world in which new possibilities for thought and action are projected have an effect on the very manner in which things are manifest in meanings. Thus an event not only effects how things look now, but also how things will look in the future, given these intimate connections of meaning and possibility. When a child is born, to use an example from Merleau-Ponty, its presence instantly redefines the meaning of everything in the world; the newborn is the introduction of the world to a new subject, a new set of concerns which, in the course of a life beginning with the minute, the hour, and the day, discovers over and over the issue of concern itself, thereby rendering a new course of possibilities suddenly, unexpectedly concrete.

This means that a reflection on the meaning of science is also a reflection on what is possible, on the possibilities not simply of science as one human activity among others, but science as a possibility of humanity, understood as a form of life.

Up to now, I have only indicated, for the purposes of providing a preliminary orientation for the interpretation of Husserl's *Crisis* given below, certain general features of Husserl idea of a reflection on the meaning of science. This does not yet determine the specific manner in which reflection can approach the claim of science, nor what this claim will look like once reflection evolves into critique. It will be argued in the next chapter that Husserl approaches this problem by way of a consideration of contemporary motivations for a reflection on science. For in Husserl, reflection, too, is motivated; it is not an activity from nowhere that opens the philosophizing subject to being motivated by the unity of sense that reflection reveals. Reflection is grounded in an already revealed concern, a question that makes itself felt within the situation of a life for which science is something "established," however problematic. But the real motivation for reflection is no longer a formal issue, but has to do with what Husserl calls the "crisis" of science, which is the subject of Chapter One.

Notes

[1] The fact that the Crisis is an historical introduction may distinguish it from other published works, but this is not a strategy Husserl adopted only in the 1930's, as his lectures from the early 1920's posthumously published under the title of *Erste Philosophie* (Hua VII-VIII) demonstrate.

[2] As, for example, in the *Logos* article (1911): "The spiritual need of our time has, in fact, become unbearable. Would that it were only theoretical lack of clarity regarding the sense of the 'reality' investigated in the natural and humanistic sciences that disturbed our peace [...] Far more than this, it is the most radical vital need that afflicts us, a need that leaves no

point of our lives untouched." HSW 193; Hua XXV 56. Also see the articles on "Erneuerung" Husserl wrote for the Japanese journal Kaizo in 1922/3 (Hua XXVII 3-42).

[3] Not that either strategy is illegitimate. For an interesting contribution along these lines, see A. Banfi, "Husserl et la crise e la pensée européenne," in: *Husserl.* Cahiers de Royaumont III (1959) 411-427. For a more general introduction to Crisis, see Aron Gurwitsch, "The Last Work of Edmund Husserl," in: *Philosophy and Phenomenological Research* XVI (1956) and XVII (1957).

[4] The use of the term *Besinnung* and its variants (*Selbstbesinnung*, *Rückbesinnung*, etc.) are ubiquitous in the *Crisis* as well as other texts of this period. Cf. Crisis 17-18, 58-59, 71-73, 335-341, 389-395; Hua VI 15-17, 58-60, 72-74, 269-276, 508-513; Hua XXIX 40-41, 192, 266-267.

[5] David Carr draws attention to an interesting distinction between *Besinnung* and *Reflexion* in Husserl's texts, where the former signifies the personal, existential preparation for the latter, which is identified with philosophy proper. See Carr, *Phenomenology and the Problem of History* (1974) 60; cf. Crisis 392 n; Hua VI 510 n, which Carr cites.

[6] See in particular the beginning of C16 I.

[7]As in FTL 10; Hua XVII 14: "Radical sense-investigation [*Besinnung*], as such, is at the same time criticism [*Kritik*] for the sake of original clarification." The deep connection between reflection and critique is of course not unique to Husserl. In this connection see Dieter Henrich's study on the conception of deduction employed by Kant, in which "reflection," or the free articulation of something the sense of which we are already in possession, plays a key role as the preliminary step of criticism. D. Henrich, "Kant's Notion of a Deduction and the Methodological Background of the First Critique." In: *Kant's Transcendental Deductions* (1989) 29-46.

[8] That is, insofar as a growing unease with the obscurity of fundamental concepts is essential to the development of science. *Sein und Zeit* 9: „Die eigentliche ‚Bewegung' der Wissenschaften spielt sich ab in der mehr oder minder radikalen und ihr selbst nicht durchsichtigen Revision der Grundbegriffe. Das Niveau einer Wissenschaft bestimmt sich daraus, wie weit sie in einer Krisis ihrer Grundbegriffe *fähig* ist."

[9]Cf. FTL 9; Hua XVII 13: "Sense investigation [*Besinnung*] signifies nothing but the attempt actually to produce the sense 'itself,' which, in the mere meaning, is a meant, a presupposed, sense; or equivalently, it is the attempt to convert the 'intentive sense' [*intendierenden Sinn*] (as it was called in the *Logische Untersuchungen*), the sense 'vaguely floating before us' in our unclear aiming, into the fulfilled, clear, sense, and thus to procure for it the evidence of its clear possibility." Commenting on these passages, Suzanne Bachelard (*La Logique de Husserl* (1957) 29) argues that "prendre conscience [*Besinnung*] c'est justement faire passer le jugement qui est simple opinion à l'état de jugement rempli." This is perhaps more true for FTL than for the Crisis; yet even in the former, phenomenology is not simply the clarification of science, there is instead a movement: the clarification of science leads to phenomenology, or the problem of the clarification of philosophy. Cf. Hua XXIX 27-36 (Text Nr. 3, "Die Naivität der Wissenschaft").

[10] Cf. Hua XXIX 3-17 (Text Nr. 1, "Menschliches Leben in <der> Geschichtlichkeit"), also Hua XXIX 41:15-46:15.

[11] Cf. Hua XXIX 165: 5-26, 226: 11-25, 229:35-39 (though this is from a text marked by Husserl as "fit only for the wastebasket—227 fn 2), 230: 1-6.

[12] Cf. Wilhelm Weischedel's *Das Wesen der Verantwortung* (1933), in particular §4.

[13] Though the point is not to speak *for* onself, apart from the others, but to become the subject of a kind of universal reflection, one that shapes a common world-horizon. Cf. XXIX 226:11-25, a passage that we will turn to again in Chapter One.
[14] Again, cf. FTL 10; Hua XVII 14: "Here original clarification [*Klärung*] means shaping the sense anew [*neuen Sinngestaltung*], not merely filling in a delineation that is already determinate and structurally articulated beforehand."
[15] Cf. Ideas I 106-107 and fn., 328; Hua III,1 101:15-19 and fn., 316:5-39; Ideas II/Hua IV §§55-56.
[16] Ideas II 243; Hua IV 231.
[17] Jorge Luis Borges, "Funes, The Memorious," in: *A Personal Anthology*, ed. A. Kerrigan, Grove Press 1967, p. 42.
[18]In a text from 1934 (Hua XXIX 40:1-41:14), Husserl outlines two levels of "historicity," each of which is characterized by a mode of *Besinnung*: reflection as natural, human self-consciousness as an historical subject, and its "Umbildung" or transformation through science. The examples of "natural" reflection/reason are instructive: the reflection of an individual on the chances of life, on fate, on what is possible and what is impossible; the king or the functionary of the state, who reflects for a people, thus on the fate of the life of the nation; the elder as a keeper of memory, who is essential in the education of the "next" generation—that is, essential in giving the parameters for a reflection no longer oriented towards what has been, but to what will be. "Vorbesinnung," Husserl writes, "setzt Rückbesinnung voraus." In each case, reflection is about and for the sake of life. This does not change, though it certainly takes a different form, when historical "reason" is transformed first through science and philosophy, then through phenomenology. Cf. Hua VI 502-503 (Beilage XXVI).
[19] Rudolf Boehm, in his comparison of the "perspectives" of Nietzsche and Husserl, presents the former as the defender of "life" against reason, the latter as the champion of a "new rationalism" (*Vom Gesichtspunkt der Phänomenologie* (1968) 218). Yet one may well be surprised, above all Nietzsche himself, when Husserl in the *Crisis* progressively appropriates the one philosopher that Nietzsche "sets apart with high reverence"—namely, *Heraclitus* (cf. *Götzen-Dämmerung* (1889, 1999) 75).
[20] Leaving aside the question of Husserl's relation to the Enlightenment, there are other ways to the question of politics in phenomenology, insofar as one can pose the question of the essence of the political from within a reflection on intersubjectivity in general and the communal in particular. For a recent contribution to this debate see Natalie Depraz, "Phenomenological Reduction and the Political," *Husserl-Studies* 12 (1995) 1-17.
[21] Hua XXIX 29:26-27.
[22] Hua XXIX 48: 30-55.
[23] See Hua XXIX 32:5-27.
[24] See Hua XXIX 7:22-11.

Chapter One

The Concept of Crisis (*Crisis* §§1-7)

Philosophorum nostri temporis vox universa

—Motto of the Belgrade journal *Philosophia*

Is there a crisis?

In November 1935, forbidden to either publish or speak publicly in Germany, Husserl traveled to Prague to give a series of lectures at the invitation of the Cercle philosophique.[1] These lectures were one of the sources for Husserl's manuscript of the *Crisis*, the first two parts of which were originally published in the first volume of the Belgrade journal *Philosophia* in 1936. The title for the lectures was "Psychology in the Crisis of European Science," reflecting a two-pronged approach that can still be discerned in the final (though still incomplete) version of the *Crisis* itself: first, the demonstration of the necessity of phenomenology in light of the "historical-teleological" essence of the contemporary philosophical situation; second, the introduction to phenomenology through a discussion of the shortcomings of modern psychology as a science of subjective life. Together, these two lines of presentation are meant to show that the crisis of psychology is not an isolated problem, but is indicative of a deeper crisis of the idea of science itself; whereas phenomenology, in breaking through to a radically different conception of subjectivity, succeeds in providing a foundation for human reason where modern psychology had failed.[2]

These lectures, in short, are about the historical importance of transcendental philosophy; but Husserl does not begin with this assertion. He begins the same way he will begin his first *Philosophia* installment in the following year (1936), namely with a question, and a reasonable one: is there, in fact, a crisis of the sciences?[3] To assert that there is a crisis of the sciences, Husserl remarks, would seem to be, at best, an exaggeration: "Can one talk seriously of a crisis of the sciences, of the sciences as such?"[4] That would imply that science has failed to be scientific (*wisssen-*

schaftlich), that it has failed in securing its proper method and task. However, if anything—particularly in light of the European social and political situation of 1935—science would appear to be the only cultural institution that is *not* in crisis, instead enjoying steady progress and development. To be sure, sciences have changed, some even radically; in the *Philosophia* installment of 1936, Husserl mentions the modern revolution in physics headed by Einstein, which could be understood as a response to the "threat"—maybe even the "crisis"—represented by what Husserl in the *Crisis* calls the "paralysis of classical physics."[5] Nevertheless, science itself, the tradition of rational inquiry as such, surely *that* is not in a crisis.

There is truth in this, and Husserl is careful to point it out, perhaps a little embarrassed by the shrill tone of the advertisement for his lectures. Moreover, it is a point that can be made even in the face of all the discussions on the foundations of the sciences, above all mathematics, that had taken various forms since the second half of the 19th century. Take for example the debate between Hilbert and Brouwer on the foundations of mathematics in the 1920's: it could be argued that what really drove the debate was not a crisis in mathematics, but rather a philosophical crisis that had emerged with the necessity of reforming Cantorian set theory in light of the various paradoxes that had been discovered by the end of the previous century. Both Hilbert's and Brouwer's programs were rather conservative attempts to operate within a philosophically palatable conception of mathematics; and both programs, despite their weaknesses (Hilbert's in light of Gödel's theorems, Brouwer's in light of the fact that it is just too restrictive), still have influence today, not because they have led to great advances in mathematics, but because they express important and valuable philosophical ideas about the nature of mathematics. But mathematics itself, its advances and failures, even its methods and conventions, could be said to be more or less indifferent to these debates on foundations; or at least that the relation is not unambiguous. For example, the *failure* of a movement like intuitionism to restrict the use of certain logical principles in mathematical analysis (such as the law of excluded middle) could be cited as an important historical factor in the *positive* development of mathematics in the 20th century, an observation that could lead one to believe that the scientist is sometimes better off being oblivious to whatever foundational crisis may or may not have a grip on his or her science.

To be sure, with psychology, the science that Husserl takes to be of central importance, the situation is much more complicated, at least in part because its development was more directly tied to the fortunes of various philosophical movements, whether the neo-Aristoteleanism of Brentano or, more importantly for Husserl, an "objectivism" that takes the physical sciences as a methodological model. The apparent failure of the extension of the inductive methods of the physical sciences to the study of subjective life represents both a crisis in the foundations of psychology as well as a crisis of objectivism, which means a crisis of philosophy itself. Nevertheless, the center of gravity in science remains the physical sciences, and whatever the woes of psychology may be, science as a whole is not on the verge of collapse; the phenomena science seeks to explain have not, once and for all, proven to be incomprehensible; the methodology of science has not been found to be in principle limited or, worse, irrational; what had been accepted as scientifically true, has not been found by other means to be simply false. It could even be argued that science, however dramatic its history, has *never* faltered *as science*, as the horizon of rational human activity and accomplishment. However different they may otherwise be, in the end Aristotle and Galileo, Newton and Einstein are all participants of a unified, forward moving tradition. Perhaps what has become questionable as a result of the failures of modern psychology, objectivism, and various foundational debates is whether *philosophy* can any longer be said to play a role in the development of the sciences, or whether it has instead come up short—but that by no means represents a crisis of the sciences, just of philosophy.

In a way, even as a result of the argument of the *Crisis*, science remains an unquestionable success.[6] That its very success does not preclude the possibility of crisis is a key insight of Husserl's; but it means that to talk of the crisis of science is, paradoxically, to talk of the crisis of a success. The crisis Husserl seeks to understand and address is thus from the start ambiguous, at least to the extent to which he continues to accept some form or gesture of the scientific attitude as legitimate. Nevertheless, it becomes clear when reading the Prague lectures that Husserl clearly has the expectation that his listeners believe that there is, in fact, a crisis; but at the same time, he recognizes that this belief is compelling only if the question of science is approached from a particular angle, one that has little to do with an assessment of the current state of the sciences in light of what science intends to accomplish.

This "angle" has to do with the place, or role of science in human life. The very success of the project of science, the continuous and rapid advances in its explanatory force, is shadowed by an emerging sense of failure; this failure remains external to scientific thinking itself, because it does not signal a failure of the program as such. In a sense, it is a failure of the very success of the program of science to advance towards a goal that belongs to another project, that of "human existence":

> In our vital need—so we are told—this science has nothing to say to us. It excludes in principle precisely the question which man, given over in our unhappy times to the most portentous upheavals, finds the most burning: questions of the meaning or meaninglessness of the whole of this human existence. Do not these questions, universal and necessary for all men, demand universal reflections and answers based on rational insight? In the final analysis they concern man as a free, self-determining being in his behavior toward the human and extrahuman surrounding world (*Umwelt*) and free in regard to his capacities for rationally shaping himself and his surrounding world.[7]

In other words science, in its contemporary form, fails to address what needs to be addressed, not in order to be a science, but in order to be a human being. Humans call upon reason to make a fundamental distinction for the ordering of human life, namely, the distinction between the meaningful and meaningless, to use the language of the Introduction. The need for such a distinction is not a condition for rationality, if by that is meant a thinking that finds its expression in certain specific patterns of judgments or assertions (called "logical"), but it is a necessary condition for human existence, to which belongs thinking but which is not reduced to thinking. Human existence requires an evidence, or experience of truth richer than logical consistency, more concrete than the true proposition. "Of all the prejudices that have gained currency, the most ruinous is that evidence is the same as 'logical' evidence, that mediate evidence amounts to deductive and immediate to axiomatic evidence, or that of immediately evident assertions ('judgments')."[8] The "claim" of science with respect to what is meaningful and meaningless is thus an existential claim which addresses the need for an evidence that is more than the conditions for rationality, and in Husserl's *Crisis*, the claim of science will be considered in light of this fundamental human need for meaning and evidence. This means that Husserl's reflection (*Besinnung*) is not merely driven by a concern with what science can tell us about the world, but more importantly, by a sense of why it is that we need science to tell us what it does, what extra-logical interests and questions it addresses.

Another issue, much more prominent in Part I than in the rest of the *Crisis*, has to do with Husserl's use of language such as "burning questions about human existence" and the like. The 1930's was, of course, very much a time of such burning questions, and one might very well object that Husserl is exaggerating the immediate impact that any philosophy can have in such times. Husserl did in fact believe that philosophy was an essential cultural institution, that it had an important, even decisive role in human self-understanding, and that the philosopher had a duty to address the issues of the day, at least on some level. Yet the point being made in these pages represents perhaps not so much a foray into social commentary as the more modest philosophical gesture of arguing that any answer to such "burning questions" cannot remain within the limits of a purely theoretical, indifferent comprehension of a thesis from a distance. Such an understanding would be in this sense more than what science, at least on the surface, claims to be, given that it presents itself as a theoretical mode of understanding par excellence. Or, in other words, if the basis from which one speaks in the name of reason is one's own personal existence, then the answer of reason to the questions of existence can only arise from an understanding that participates in the shaping and forming of a way of being. Yet Husserl need not go so far as to claim that he is in possession of a program for a new way of life; one can accept this idea of a question in which how one is to be, to exist is at stake without at the same time taking for granted that philosophy can either adequately pose or answer the question. One does not need to be a philosopher to be burned by a question, nor respond to it; but to be a philosopher, in Husserl's sense, one must be burned, as well as find within oneself the resources for a response.

In turn, what it means to formulate an answer to "burning questions," or even to reflect on the situation in which such questions must be addressed, is always more than a practical problem. The question of human existence, as Husserl introduces it in the *Crisis*, can be posed only with a view to the inextricability of existence and meaning; human existence for Husserl is dependent on a sense of and for the meaningful to provide a context of thought and action in which it can unfold. This implies that a sense for the existential weight of the meaningful is operative in theoretical activity as well, even if, as in pure science, it is premised on a deliberate isolation from practical life in principle. One could even claim that, for Husserl, the highest form that this sense for the weight of meaning can take is that of an explicitly theoretical sensibility, though where the theoretician does not remain "objective," or "indifferent" to the fundamental needs of the human project, whether they manifest themselves as practical, theoretical, or a combination of the two.

It is usually as a combination of the two that such questions, and the needs they express, are experienced. Lodged somewhere between theory and practice, burning questions of human existence are practical difficulties in which the existence of and reliance on theoretical principles play a central role. There were many such questions in the 1930's, ranging from national identity and historical purpose to the struggle over the justice and future of capitalism, and they were all marked by the complex interface between ideas and conflicts in a world shaped by both. This was, after all, a decade dominated by fascism and communism, which together achieved a high level of systematization and, above all, aesthetization of political ideology and violence. Husserl thus had a lot of competition in the art of posing and addressing burning questions; and in hindsight, perhaps he was all too willing to indulge in the politically charged rhetoric of questions that "must" be addressed, of situations that have become unbearable, of "decisive moments" faced by the "will" of generations, and of great events that will forever change human existence.[9]

As proposed in the Introduction, this interpretation of Husserl's *Crisis* will not rely on situating the text historically, so the historical background of Husserl's politics will not be my concern here. Yet, the invocation of "burning questions" is more than a reflection of the political rhetoric common to the 1930's; it in fact has a philosophical core. For Husserl, to address these kinds of questions requires not only an understanding of what is entailed by a particular historical or political situation; it also requires pursuing the philosophical task of moving beyond the situation in order to formulate a kind of sense of sense, or the meaning of meaning that will provide the basis for a global understanding of the conditions of being situated in the world, in history as such. In other words, the need for an answer to the question of what is meaningful and meaningless can be traced ultimately to the *need for a world,* a world both stable and comprehensible that can serve as a foundation for a response to the question of who and what we are.

This need can be addressed by philosophical reflection only to the extent to which such reflection accomplishes a *formation* of sense, or only to the extent to which reflection matures into critique, as was explored in the Introduction. Only the formation of sense addresses this need for comprehensibility, for a "world," because only in its accomplishment is sense something experienced, or lived. To be sure, this needs to be more precise: in what way does the formation of sense in turn fashion the world? Invoking the idea of science as technology, or the knowledge of how to manipulate and construct, does not really answer the question; the very possibility of technology is dependent upon the prior accomplishment of science in its articulation of the world. Science does not only provide homo faber with instruments; it is itself an already constructed experience of the world that infuses all such instruments with their sense. The answer has to do with the manner in which the accomplishments of sense-formation are instituted, or the manner in which meaning is not simply *fashioned* but is *constitutive* of a possibility of manifestation, or world-disclosure. In the Introduction, it was argued that science must be seen as an already accomplished reflection on the world; it was also indicated that this accomplishment is not limited to the formation of sense alone, but includes an *institution of sense* in the wake of what reflection reveals to be possible. By articulating a sense of sense, science, or better the claim of science, articulates how to make a world manifest, or visible.

It is clear that what Husserl means by "world" is decisive for this line of reflection. Consider a passage from a lecture Husserl gave in May of 1935 in Vienna, at the invitation of the Wiener Kulturbund.[10] In this passage, he takes up the concept of the "surrounding world" or *Umwelt*:

> "Surrounding world" (*Umwelt*) is a concept that has its place exclusively in the spiritual sphere. That we live in our particular surrounding world, which is the locus of all our cares and endeavors—this refers to a fact that occurs purely within the spiritual realm. Our surrounding world is a spiritual structure [*geistiges Gebilde*] in us and in our historical life. Thus there is no reason for him who makes spirit as spirit his subject matter to demand anything other than a purely spiritual explanation for it. And so generally: to look upon the nature of the surrounding world as something alien to the spirit, and consequently to want to buttress humanistic science with natural science so as to make it supposedly exact, is absurd.
>
> What is obviously also completely forgotten is that natural science (like all science generally) is a title for spiritual accomplishments, namely, those of the natural scientists working together; as such they belong, after all, like all spiritual occurrences, to the region of what is to be explained by humanistic disciplines. Now is it not absurd and circular to want to explain the historical event "natural science" in a natural-scientific way, to explain it by bringing in natural science and its natural laws, which, as spiritual accomplishment, themselves belong to the problem?[11]

The distinction in this passage between "spirit" and "nature," along with the distinction between the humanistic and natural sciences which follows, is something that cannot be taken up here, apart from mentioning that it is more subtle than it might at first appear. In Husserl's writings, "spirit and nature" is a distinction drawn within the world of experience, which in part implies that the reality of the world cannot be articulated in terms of an homogenous ontology, but must instead be divided into "regions," where such divisions by no means exclude important interrelations between regions, even dependencies. Furthermore, the ontological laws that govern these relations remain obscure until a level of analysis has been reached that corresponds neither to the ontological region of spirit nor to that of nature, but to that of lived experience, the elaboration of which inaugurates a new perspective on the problem of the world over and above the ontologies of its different regions.[12]

For my purposes here, there are two points that can be made from this passage. (1) The description of the world as "spiritual" indicates that, in a fundamental sense, the world of experience has the sense of being generated or fashioned, or in other words, that spirit marks its world with a sense and a coherence. The world of the Greeks could have been what it was only had the Greeks lived through it the way that they did. However, the point is not that the Greeks created a world, or even re-created a world they had found; the idea is more general, namely that the manner in which things take shape in experience always carries with it a sense of accomplishment characteristic of the mode in which things can be encountered. It is only from within the movement of an accomplishing experience that a subject can be aware of, and in this way encounter, anything at all; the dynamics of structuring, assembling, combining, holding-together, forming, even analyzing wholes is the very manner in which the world unfolds as a unity of meaning.[13] To be actively engaged with things, even to simply navigate among things, is to be directed towards how the world, progressively taking shape in an activity that establishes an order, manifests itself; it is to be attuned to what will result from the meaning or understanding set into place by an event, or what the very event or "happening" of an experience of the world is poised to *produce*. Events always, more or less, result in something. All expectation, for example, is grounded in this fundamental understanding of the world as result, as accomplished being in this sense. Even that aspect of the world which lends itself to a pure thinking or understanding is not excluded from this primordial accomplishing, as were the objects of Aristotle's *noein*; whatever is seen for what it is, in Husserl, is seen only thanks to the accomplishment of the experience in which it is manifest.

This results in a certain double meaning of the phrase "surrounding world." Take for example the expression "spiritual structure" in the third sentence of the passage quoted above, which David Carr in his translation of the *Crisis* uses for the German phrase "*geistiges Gebilde.*" Husserl often uses the word *Gebilde* in texts from this period, and part of what makes it attractive is that its connotations are not as abstract as with the word "structure"; namely, *Gebilde* can simply mean "thing." This is in fact important to point out, since on one level Husserl is merely saying that the surrounding world is a world of "things"—things that have meaning or significance precisely in the sense of things "in" and "of" this world. But *Gebilde* also means "thing" in the sense that a formula or a theory, a custom or a poem by Montale is a "thing"—thus precisely in the sense of a structure that can "be" at all only insofar as it has been moved, shaped, or fashioned by a concern that has led to or resulted in its accomplishment. In lived experience, the being of a "thing" does not rest on itself alone; a thing is never encountered as it would encounter itself, "in itself"; yet nor does the being of a thing rest on other "things," but on a context of things defined by the movement of an experience directed at things as its accomplishments.[14] This need not yet be taken as a thesis of a full-blown transcendental philosophy, even though for Husserl its full implications do in fact require a transcendental reflection; it can, initially, rest solely on our common subjective familiarity with the various styles in which our lived experiences unfold.[15]

"Reason" as the capacity to form and shape the world can be identified as a mode of this subjective accomplishment, thus as the basis for a particular manner in which the world is manifest to consciousness. Alternatively, in the language proposed in the Introduction, the possibility of being in the grip of a meaning is conditioned by the capacity to be the subject for whom the world progressively takes shape in the accomplishment of an understanding of it.[16] Thus the result of emphasizing the concept of world as "Umwelt" is not to suggest that the world is a "mere construct," but that, for the world to have meaning in the mode of being what it is in experience, it must be brought into play within a subjective movement—"constructed" in the way that an idea is constructed by the act that expresses it, or *King Lear* by the event of its performance.[17] This point is simple, but critical, and plays a key role in Husserl's reflections. The rather dubious idea of affirming the existence of the world "itself" as pure transcendence beyond any subjectivity should not be confused with the intuitive sense of concreteness or objectivity; giving up the philosophical myth of the "objective world" does not undermine the coherence or place of the latter. The intuitive sense of the concrete refers to the embodiment of the world in experience; it characterizes that givenness of self which the world can assume only from within the activity of lived understanding, thus within its construction in accordance with a form and idea that belong to the particular articulateness of a lived experience. Abstracted from this embodiment, the world is merely an "it" without a self, an object without objectivity, a *tode ti* without a "what." This means that the question of the world as lived is the question how the world has become embodied in the way that it has—or how spirit accomplishes precisely *this* meaningful coherence of world, *this* signification that has become what the world is "itself."

In light of this, (2) science for Husserl is not a picture of the world, an interpretation of things that may or may not be accurate, but in a deeper sense it is part of an experience in which the world is embodied.[18] Science itself, as Husserl states, is a spiritual accomplishment; it, too, belongs to the "surrounding world," as the (successful) formulation of an understanding of the things encountered in it. It is the result of the exercise of the capacity to be open to a manner in which the world can take shape in the understanding. Even if the natural scientific understanding of the world takes great pains to understand its object "as it is," independently of anything subjective, such an understanding remains an accomplishment of the human spirit. There is no circularity here, only an example of that irremovable double meaning thing/construct. What is understood thanks to the natural sciences, the purely objective world, is itself part of the coherence of produced meaning that belongs to the "surrounding world" (*Umwelt*); but it is no less true that what is thereby understood is simply the world itself, or the objectivity that results from the formative event of understanding. Science, as a spiritual accomplishment, a successful reflection on the world, has projected the possibility of an experience of a world understood precisely in the sense of that which is "purely objective."[19]

Thus when Husserl speaks in Part I of the *Crisis* of the success of the sciences, this refers to the fact that science has successfully articulated an understanding of the world; and in doing so, *it has become, to a great extent, determinate of the world in which we live*. The reason why our relation to science is so complex is because, as rational beings, we are unable to hold it at arm's length; we cannot easily compare the world of our life to the account of it given by science, because science is also operative on the very level of world-disclosure itself. Anything less could not be called an accomplishment of an understanding, of the spirit. For what does understanding, as an act, seek to do? For Husserl, the answer can only be the accomplishment of sense, which is again not merely a picture or even an interpretation, but an openness towards a particular mode of manifestation. Even the interest in the abstractions of mathematical logic ultimately rests on an interest in the world, the world as it is; the idea that we seek out abstractions for their own sake, out of some innate curiosity of a reason easily seduced by the fairy tales of the non-real, is for Husserl nonsense—genuine thinking, as with all dealings and directedness towards the sphere of meaning, is possible only in the wake of a world-interest.[20] And it is in the interest of living in the world that we seek to understand it—because it is only as an understanding of the world that the subject is "in" the world at all.

The accomplishment of understanding thus does not simply present or conceive a truth, but places the subject, so to speak, in its presence, in its midst. Science, in other words, articulates the truth of the world in which we live: when we want to get closer to things, to understand them better, the claim of science determines the horizon of our approach. But what sort of "truth" is the truth of science? Correlatively, what sort of "world" is its correlate? Husserl in the Introduction to the *Crisis* invokes a traditional, even everyday conception of science: "Scientific, objective truth is exclusively a matter of establishing what the world, the physical as well as the spiritual world, is in fact."[21] The idea of a "fact" belongs to the content of the claim of science about what is to be considered rational/meaningful with respect to the world as such; more importantly, it determines what types of questions can be posed: no question that does not bear, directly or indirectly, on a question of fact is meaningful, nor is any answer to a question of fact that itself does not arrive at what is in fact the case. We can call this objective perspective of science "factological," insofar as what we mean by that is the claim that the *logic of the world* is a *logic of facts*.

What Husserl means by "fact" is something localizable in the objective world, like an object in space and time or an event within a nexus of cause and effect.[22] It refers, in other words, to a kind of purified vision of things, one that is arrived at by way of a process of exclusion. The story is familiar: in science, everything "merely subjective" is excluded in favor of what is "purely objective," or what belongs to the world disclosed independently of any colorings of the subjective. This process—which is again a mode of constructing, or even re-constructing—allows the world to appear in understanding "on its own," in accordance with the form of independence or autonomy from that which is subject-relative.

The orientation towards such a purified, objective givenness is the goal of what Husserl calls "objectivism." He uses this term in a pejorative sense; yet it should be emphasized that the impetus behind objectivism is again the very success of the articulation of the logic of the world as fact within the physical sciences. Objectivism seeks to extend this success into all spheres of human understanding, in part by cultivating a sensitivity to the demand to exclude the merely subjective as irrelevant with respect to an understanding of "reality." Husserl is convinced that such an extension ultimately undermines itself by resulting in a series of paradoxes, above all with respect to the problem of the nature of subjectivity itself (psychology); but in a way, Husserl's philosophy itself depends on the same sources as objectivism, namely the cultivation of an ability to carefully identify the presence and contribution of the subjective in what we see and experience. Husserl's opposition to objectivism turns on the insistence of the latter that the subjective be devalued, that all value is either objective value, or valued in reference to the standard of factual objectivity.

Objectivism and its devaluation of the subjective is more than just a theoretical project, it is a tendency within modern life itself, namely the tendency to orient understanding around those features of the given surrounding world that are compatible with the understanding of the world as objective. The result is that the surrounding world has become, increasingly, an "objective" world, understanding an understanding of fact; but that can take place only thanks to a kind of reductionism that severely *limits* what can rightfully assume the appearance of the meaningful. This results in a kind of radicalized world, one in which even human existence itself appears not to have a legitimate claim to being counted among those things that are meaningful:

> But can the world, and human existence in it, truthfully have a meaning if the sciences recognize as true only what is objectively established in this fashion [that is, as "fact"—JD], and if history has nothing more to teach us than that all the shapes of this spiritual world, all the conditions of life, ideals, norms upon which man relies, form and dissolve themselves like fleeting waves, that it always was and ever will be so, that again and again reason must turn into nonsense, and well-being into misery? Can we console ourselves with that? Can we live in this world, where historical occurrence is nothing but an unending concatenation of illusory progress and bitter disappointment?[23]

The radical orientation of science, the claim of science in its unfolding as a tendency of modern life, restricts the scope of meaning or sense to the point where it no longer serves as a compelling mode in which the order of life and world is articulated. Too many of our beliefs, too much of our history, and with that too much of our existence has lost the status of "meaningful." Too much of the world we have come to understand, thus to inhabit in a particular fashion, has become "meaningless" as a result of the orientation of our life. And this is not only because science has encountered something in the world that it cannot explain; the crisis of the sciences is more than the sum of theoretical difficulties in adopting a reductivist/objectivist perspective in particular areas of research such as psychology. For Husserl, these difficulties point to a deeper life-crisis of European humanity, one that has led to the uninhabitability of the world, not because there is something that cannot be understood, but because in objectivism, understanding itself, at the height of its influence and power to shape the surrounding world, has lost its footing, giving rise to an existential crisis about the possibility of the very decision it represents for human existence—the decision about what is to count as meaningful and meaningless. The employment of the very means by which humanity is to find its place in the world has led to its exile.

Is there, in view of their constant successes, a crisis of the sciences? By the end of Part I of Husserl's *Crisis* this question is in effect replaced with another: *when did the world become incomprehensible to us*? The success of science is left intact; the question instead has to do with the implications and effect this success has had on a humanity that is dependent on its understanding in order to live. Science is successful because it restricts itself to the "objective"; the systematic, methodical purification of the understanding has led to an explosion of developments in the sciences, in the wake of which a world (*Umwelt*) has been fashioned. Yet the restrictions emphasized by an objectivism that seeks to articulate the sense of this success are too severe, its language too alien to be able to address humanity in a way that can be comprehended; as human beings, we do not understand what science says when it speaks to us, thus in an important sense what the world says to us through our experience of it through science. Neither world nor reason seem to have "anything to say to us," which means that, in the end, the need for a world is left unanswered, even if it is indisputable that we have understood the world, thus possess it in a genuine, concrete sense—that, however alien it may otherwise be, it is "ours." Moreover, this question of the incomprehensibility of the world is identical with the question, itself a "burning question" of human existence: *when did our world become uninhabitable*? When, and because of what did we lose our home? This, in short, is the double question of the "crisis of European sciences."[24]

"Vernunft wird Unsinn, Wohltat Plage"

In his translation of the *Crisis*, David Carr points out that the phrase "reason must turn into nonsense, well-being into misery" in the passage cited above is a quote from Part I of Goethe's *Faust*. The line from Goethe runs "Vernunft wird Unsinn, Wohltat Plage," which Walter Kaufmann translates in a suggestive way: "wisdom becomes nonsense; kindness, oppression." It is worth reflecting on the scene from *Faust* in which this line appears, because it can serve to introduce in a preliminary way the way that Husserl will develop the question of the origin of the crisis as an historical phenomenon.

In the scene in *Faust* in which this line appears, Mephistopheles, impersonating the despondent Faust,[25] is conducting an interview with a student who is trying to decide which course of study he should pursue. After Mephistopheles, somewhat to his irritation, fails to seduce the student with logic and metaphysics, the latter confesses that he is not much inclined towards jurisprudence either. To this Mephistopheles responds (again using Kaufmann's translation):

> For that I cannot blame the students,
> I know this science is a blight.
> The laws and statutes of a nation
> Are an inherited disease,
> From generation to generation
> And place to place they drag on by degrees.
> Wisdom becomes nonsense; kindness, oppression:
> To be a grandson is a curse.
> The right that is innate in us
> Is not discussed by the profession.

Mephistopheles' complaint turns on a difference between the law that is innate, and the law which is practiced, where practices are not only the individual activities of persons but take the form of inherited traditions. The contrast is a familiar philosophical gesture. The "right that is innate in us" takes the form of a feeling for lawfulness and right that is independent of the "laws and statutes of a nation." This independence also cuts the other way, however, in that the study of jurisprudence, or the study of the tradition of law, can be exhaustive from a traditional point of view but still overlook the innate sense of right. This also means that the tradition, or "traditional right," which exists as a part of the world in the manner of a particular set of instituted practices, takes on certain characteristics in light of the particular history that shapes it qua tradition. There is a difference between what the world has because of its history and what the subject has because of its nature. And though these two sides are in some sense independent, it is not as if they had nothing to do with one another; there is always a need, or a demand, however obscure its formulation, that the innate sense of right be satisfied with whatever historical or traditional manifestation justice actually takes. Not that our innate sense is necessarily present as a clearly articulated standard or norm; it is enough to be aware merely of the difference between the ideal and the real, and to understand that whatever may begin as something close to the ideal (wisdom, kindness, good intentions) more often than not turns, given enough history, into something closer to its opposite. "To be a grandson is a curse"—cursed to be at the mercy of the lingering trespasses of the past.

There is a similar distinction between the ideal and the real that governs Husserl's conception of the crisis of the sciences. The articulation of the crisis is guided by a sensitivity to the difference between the ideal of science and its historical manifestation, i.e. the results of its having been "worked out" in concrete historical life. To be sure, the description of this difference between the ideal and the real, in particular with respect to the use of the contrast between "inner" and "outer," will be very different in the *Crisis* than it is in this passage from Goethe. Though the "idea" of science will be something "inward" for Husserl as well, it will be an inwardness or immanence within the activity of science itself, in the sense that any activity contains within it some prior sense of its idea, however obscure that may be.[26] The basic intellectual gesture, however, is very similar: the crisis of science is genuinely grasped from the perspective of the question of the origin of science—an origin which, like the wisdom of the grandfather, is putatively closer to a more pure, a more "ideal" sense of what science is "supposed" to mean, or to intend as its goal.[27] Moreover, because of this proximity to the ideal, this original intention of science is to be distinguished from what happens when it is taken up and lived by a culture.

This passage in Goethe also leads to another issue concerning the role of the "ideal of science" in the *Crisis*. The innate sense of right is, as innate, implicitly timeless; what comes innate is not passed on as something inherited, which means that every given historical subject has an equal share in the resources implicit in the innate. In turn, an intuition of the idea of science is not dependent upon an institution of a standard at an initial point in history in which the law was "born," but can also be understood as an originary motivating factor that functions throughout the historical tradition taken as a whole. The law as "idea" is therefore a continuously present source of motivation, and not simply the preserved memory of an assertion or conviction once held by those who began the tradition. The innate sense of right thus represents an "origin" that can be understood as a constant accompaniment of the tradition, one that lays claim to the right to define the future movement of the tradition as an historical phenomenon.[28] Something similar will be the case with the idea of science in the *Crisis*: it, too, will be an "origin" accessible throughout the tradition, though its accessibility will not be secured by the non-traditionality of an "innate sense."

Yet, the presence of the idea, whether or not it is secured by an innate sense, is not enough to establish its influence on the real. This is not only a question of accepting or rejecting the idea, but above all of the concrete articulation of what the idea "means." There is always a difference, constantly felt or at play, between the idea as motivating factor and the development of a sense for the concrete implications of the idea in history. Historical becoming continuously fashions the possible, setting it up in different ways, and that is no less true with respect to the possible senses of an ideal. However timeless it may be in itself, Goethe's innate sense of right is, for those who possess it, nevertheless marked by what has happened to it in time—by its neglect, re-interpretation, subjectification, marginalization; by all the decisions made in its name, and all the exceptions that have compromised (or enhanced) the force of its validity.

Likewise, the originary development of scientific culture, in the wake of its idea, opens a world of possibilities that had been dormant or nonexistent prior to the original intuition of the idea as the event of original institution, in which the idea first becomes a motivating factor in history. And these "new" possibilities, like any possibilities, are concrete only to the extent to which they are to be acted upon, which means above all that decisions must be made about what they are going to "actually mean." Such decisions are not always a simple matter; and they do not always lend themselves to being made simply by following the guidelines of an ideal. On the contrary, the ideal itself, or the manner in which the motivation of the idea actually plays itself out, is one of the things being decided when faced with a possibility of interpretation. Thus, gradually, the sense of the original ideal of science itself begins to change, it begins to be discovered, or to motivate, from within the perspective of a concrete tradition, where its "original" sense is relativized within a multiplicity of possible articulations. One of Husserl's most important insights in the *Crisis* is that it is only within this unfolding "reality" of the meaning of science that it exercises its hold, for it is only as an actual *historical phenomenon* that the idea has any *force*. This means that the critique of the idea of science, as a *Besinnung*, is for Husserl necessarily an historical reflection, not in order to trace a traditional inheritance back to an original, pure idea, but to uncover the historical possibility of the origin taking hold as such—again. Husserl's appeal, therefore, will not be to recognize an "innate" grasp of the idea, but for an encounter and appropriation of the latent possibilities of a tradition.

The crisis expressed in the tendency of objectivism is also an historical phenomenon, which means that it owes its force to its concretization within a tradition. Objectivism is a tendency within a concrete articulation, thus realization of the idea of science, and as such represents, in principle, a kind of risk; for realizations always put at risk original intentions. This is not only because of the possibility that the ideal will not be realized but, more importantly, because what is to be in turn possible for the future, the form that possibilities themselves will take on as "real" or "actual," is also being shaped. More, this risk goes beyond problems that arise due to the specific limitations and restrictions of objectivism; the deeper question is what has been closed off once these restrictions of objectivism tend to be accepted, that is, once they no longer represent provisional experiments with the idea but have become a tradition of successful practices. What is at stake when a tendency like objectivism takes hold in the context of a developing history is the very openness of possibility. The rise of objectivism represents a phase of the historical development of what is possible in the wake of the idea of reason, and as such, it initiates the necessity of the preservation of the possible beyond the limits of what it itself has made real. Yet, and this is an insight that Husserl gradually develops in his reflections between 1934 and 1937, this would be true of *any* historical realization of the idea, and cannot be attributed to the consequences of objectivism alone. Every historical possibility of understanding science, originally opened by the intention of the "first" scientists, also necessarily opens up the possibility of misunderstanding, though in new and different ways.

This is again not unlike what Mephistopheles is complaining about when he talks of the "laws and statues of a nation" as an "inherited disease," a "curse." A law in its original institution is a living decision; it is a first step, a new projection, a pure gesture, a first encounter. However, in its traditional form, a law is a decision we inherit. As such, the law has already shaped certain possibilities with respect to a particular question (again by way of a process of exclusion: some are rejected, others accepted, some left undefined). As a decision, a question that has been answered, the law defines how a certain horizon of possible events in human social life are to be understood, encountered, and "judged." Nevertheless, even if we can call such a law or decision established, it does not, in connection with each and every actual event falling within this horizon, decide *for itself*. It is always up to someone to bring the law to what it is going to actually mean in a particular case, or what it is going to mean "for us," for those who are not the fathers of the original decision but at best the grandsons, as it were. In effect, even if what we decide does not fall outside of the well-defined constraints of legal tradition, we must nevertheless still decide again, and in making such decisions we have an impact not only on the particular case, but on the meaning of the definition or law itself, on the horizon of possible comprehension that had been originally put into place by and as the tradition. For however established, this horizon is not immutable, since we must in every case make it our own. Thus the history of the law is the history of what the "idea" of the law has become as a concrete, historical phenomenon, with all of its layers of reinterpretation, rephrasing, shifts of emphasis, forgetting, willful misinterpretation—in short, everything that accompanies real ideas, real principles, the very reality of which means that they are never pure, first encounters.

Which implies that, right from the beginning, right at the point at which the world is engaged as a context that meaning has made possible "for us," thus as something to be shaped and formed in its significance, there is a "crisis."

A brief history of the concept of crisis

Husserl was of course not the first to make use of the concept of "crisis," though the *Crisis* is an important, and in many ways unique contribution to the history of the concept.[29] The use of the term throughout the *Crisis* is for the most part commensurate with the ancient medical sense of the term, or with Augustine's use in the *Confessions* of such phrases such as "crisis of faith."[30] In traditional medical terminology, a "crisis" marks a turning point in the course of an illness: it is the point where it is going to be decided whether the patient lives or dies. A crisis is thus dangerous and decisive at the same time, and the word can be used to express the danger, or the risk, inherent to a decision. It is also an experience of necessity: a crisis is a situation where we can go no further, or carry on no longer, without a fundamental change; for better or for worse, in a crisis a decision *must* be made, it is a danger that must be resolved.

The word itself is Greek (*krisis*), and can mean: (a) a division, or a conflict; but also (b) a decision, or a judgment. Both senses can be operative at the same time, as when one refers to a judgment that can end a conflict, or which takes a final position within a conflict without necessarily bringing it to a close. *Krisis* is derived from the verb *krinô*, which means to pick out or to choose. All of these senses find their place in the medical use of the term: at the moment of "crisis" the patient is divided between life an death, "one foot in the grave"; the moment is inherently unstable, but it is also the "decisive moment," or the moment that will choose life or death.

Yet medicine is not the only sphere of life in which this word has been traditionally used. It has also found its way into politics, and very early on. And in the political use of *krisis*, particularly in the case of Aristotle, the same meanings are at play, but with a somewhat different effect, given (in part) the difference in subject matter. In particular, in Aristotle there is an interesting double sense of *krisis* that only here takes on an important role. Take for example the passage at *Politics* 1275a20,[31] where Aristotle defines citizenship as participation in both *archê*, which can be translated by "authority" or "dominion," and *krisis*, which here means of course "decision"—the citizen takes part in political decisions. There is also a natural reference here to conflict, insofar as political decisions are occasioned by conflicts, great or small; authority defined without reference to conflict is meaningless. Thus, participation in politics means that the citizen takes part in decisions that bring about resolutions, or, in other words, faces the task of choosing, in the sense of *krinô*.

It is precisely here, with respect to the issue of political choice, where this word takes on a curious double meaning. It can mean the task of decision itself, the burden shouldered; but it can also mean the circumstances that brought about, or called for, the shouldering of the burden. Thus the act of deciding need not be understood as coupled with a simple reference to a conflict or occasion, as in the case of a bureaucratic decision that appears external to the situation that required it ("in reference to case number 000 the following decision has been made"), but can instead be understood from within the concrete unity of movement that belongs to the situation itself: "I am engaged with a problem about which I must make a decision"; "I am concerned with a question that must be resolved." And for Aristotle political crisis is not just one concern or situation among others, but is an involvement that defines citizenship as such. Thus, presumably, for Aristotle the citizen is not only defined as the one who has the function of making decisions, but also as the one who is *concerned with* whatever question or conflict requires the *polis* to make a decision. That is, the citizen is the one for whom it is necessary that a decision be made, just as much as it is the citizen who makes the decision.

What is interesting here is not so much the specific etymology of the word—which in Husserl's text, it should be stressed, plays absolutely no role—but something that seems to be missing in Aristotle's text, namely, what we could, following Reinhart Koselleck, call the modern distinction between objective "crisis" on the one hand, and subjective "critique" on the other.[32] Whether or not to emphasize this distinction is, I would argue, an important question in understanding the place of Husserl's text in the history of discourses of "crisis," as well as in understanding its argument.

First the distinction, which is apparently missing in Aristotle, or at least not emphasized: critique, as opposed to crisis, need not be understood as anything more than a reflective assessment of a situation in which the critic is either not involved at all, or is at least not involved while assuming the mantle of critic. The critic, to be a critic, need not be the one who assumes the *authority*, understood in the modern context as the *right*, to a make a decision.[33] Critique as such, mindful of a (modern) sovereign who ultimately has power over anything that would be a factor in the exercise of power in the public space, does not amount to any real, objective decision—thus does not represent being in, or being concerned with, and above all not *charged with* a "crisis."

Such an opposition, and the evasion that underlies it, is apparently absent in Aristotle. To be sure, citizens exercise critique, they make circumspective judgments about what is just and unjust; but such critique is always at the same time a participation in the actual institution of justice from within the actual development of crisis.[34] Thus in Aristotle we see no need, at least not in the political sphere, to distinguish rigorously between *kritikê* and *krisis*: they belong to the same family of meanings that stem from *krinô*, to choose. There is an absence of the modern motivation to distinguish between *one who chooses* and *the one who has to make a choice*.

Again, this history is interesting, only because it enables us to formulate a question for interpreting Husserl's text: what is the relation between "critique" and "crisis"? Is it guided by the modern distinction between the two, or does the participation in one necessarily entail participation in the other? And if "reflection," understood as *Besinnung*, is in a sense identical with critique (as was argued in the Introduction), then how does this bear on the relation between "reflection" and "crisis"?

Two readings

There are at least two possible readings of Husserl's text in response to these questions. The first is oriented around the medical or therapeutic connotations of the word "crisis." Accordingly, the crisis would be either the breakdown or the threat of a breakdown of the normative function of reason, the fall into what Husserl describes with dismay as the "skepticism" haunting modern times.[35] Skepticism in its modern form is the suspicion that reason has nothing to do with life, that its project has nothing to do with who and what we are, and where this suspicion has spread to the culture as a whole, leading to a collapse of the "belief in reason." According to such a reading, the crisis of reason would be understood as a growing danger to the viability of the European sense of self, and with that of the meaningful coherence of its world, all of which is the result of the wayward development of objectivist modern science since the Renaissance. In the Renaissance, the expression of the ideal of science somehow falters; the highest ideal of the rational no longer functions effectively in the ordering of the human lived world. As a result, the original, genuine sense of Europe as a rational culture has been "lost"; the current generation has failed to live up to its responsibility of preserving the ideal that had been assigned to it and to those that had come before it.

The task of Husserl's project, according to such a reading, would be to set things right through a radical reformation of the self-understanding of science, the institution within which the idea of reason is articulated for Europe. This would be nothing less than a call for a new Renaissance, one intended to supplant the ideals of the old Renaissance with a new understanding of the idea of reason. The task would be to make the importance of reason for life "believable" again, to make the conception of a rational world, the (for Husserl) "European" world *par excellence*, once again an inhabitable world. And it would do this by bringing the European spirit once again into contact with the original sense of the idea of science, something the history of scientific culture itself has obfuscated, thereby making the possibility of shaping the world from within the horizon of reason once again a living possibility.

For such a reading, the modern distinction between "crisis" and "critique" highlighted by Koselleck could perhaps prove to be quite useful. "Crisis" would be the objective, historical situation of a rational ideal that, firmly instituted as tradition, no longer orders the world in a meaningful way; the very discourse of reason having led to a world with which we have long since ceased to be able to identify ourselves as citizens. "Critique," and the reflection that makes it possible, would be a name for philosophy itself, a philosophy that both provides a diagnosis of this situation and reveals a hitherto unanticipated depth and resource of reason, belying the skepticism that reason no longer has "anything to say" with respect to the burning questions of who and what we are—questions which, as was suggested above, come down to the question of what is meaningful and not meaningful. And philosophy would be in a position to do this, precisely because of the relative independence of critique with respect to crisis: critique is the calm, intellectual penetration into what is possible and not possible, where assertions are not really decisions, but descriptions of what kinds of decisions would or could find a place and a function in the situation of crisis. The philosopher would be like the doctor giving recommendations, formulating diagnoses and dispensing dietary suggestions to the patient, who in turn would be the one at a crossroads, a turning point where a decision must be made. The ultimate purpose of Husserl's *Crisis*, therefore, would be to provide a spiritual therapeutics for a European culture facing an immanent collapse of its own ideal, one that would take the form of a reflection, in the sense of a *Besinnung*, on the idea of reason.

This reading would be consistent with many of the texts that make up Husserl's early work on the *Crisis*, in particular the lectures to the Wiener Kulturbund in May of 1935.[36] But the development of Husserl's reflections after the initial phase of 1934-1935, I would like to suggest, indicate another possible interpretation, one in which the philosopher does not appear as the master of critique alone, or as a cultural physician identifying and treating the afflictions of the times. This alternative reading would instead be guided by the idea of an historical *Besinnung* in the sense indicated in the Introduction above, where the issue of motivation and the "hold" or "grip" of meaning plays the central role. The result would be that the philosopher looks more like Aristotle's citizen, who addresses what is at stake, thus who decides, but who also has a stake in the crisis; and where it is from the perspective of the latter that the possibility of the former must be identified. The resources that enable us to care, to be concerned for the city, are the same as those that enable us to make decisions on its behalf. Thus critique itself, according to such a reading, would be possible only for one who has taken responsibility for what has been uncovered in reflection, which means "being in the grip" of a "crisis." This grip of meaning would be the grip of a crisis of meaning, the idea being that it is perhaps originally as a "crisis" that meaning can have the grip that it does; thus only for those who believe, is something like a decision of belief possible. Such a reflection would thus not be limited to a diagnosis of the crisis, but on the contrary its intensification and, above all, its very manifestation.

To illustrate, take the following passage from Part I of the *Crisis*:

> A definite ideal of a universal philosophy and its method forms the beginning; this is, so to speak, the primal establishment of the philosophical modern age and all its lines of development. But instead of being able to work itself out in fact, this ideal suffers an inner dissolution. As against attempts to carry out and newly fortify the ideal, this dissolution gives rise to revolutionary, more or less radical innovations. Thus the problem of the genuine ideal of universal philosophy and its genuine method now actually becomes the innermost driving force of all historical philosophical movements. But this is to say that, ultimately, all modern sciences drifted into a peculiar, increasingly puzzling crisis with regard to the meaning of their original founding as branches of philosophy, a meaning

> which they continued to bear within themselves. This is a crisis which does not encroach upon the theoretical and practical successes of the special sciences; yet it shakes to the foundation the whole meaning of their truth.[37]

This is effectively a summary of the whole problem of the crisis. Notice that at no point is the argument that the ideal of science was ever *non-problematic*. On the contrary:

> [...] in its [i.e. science's—JD] first, original establishment, ancient philosophy, it conceives of and takes as its task the exalted idea of universal knowledge concerning the totality of what is. Yet in the very attempt to fulfill it, the naive obviousness of this task is increasingly transformed—as one feels already in the ancient systems—into unintelligibility.[38]

This unintelligibility is not a complete, unambiguous "breakdown" or "dissolution" of the idea of reason; nor is it simply an error that could have been avoided, a mistake that could have been fixed before it was too late. Husserl in these pages is pointing to another kind of breakdown that is not an exception, but an ever-present possibility and partial reality of a life in reason, a permanent risk inherent to any form that the realization or fulfillment of the meaning of science can take in life. The idea is that breakdown and dissolution are not just distant possibilities, but risks that must be addressed at every turn; they are real possibilities of a crisis that is intrinsic to a life lived in a world opened by the ideal of the rational, and have always already made inroads into the meaning of rational activity, always already announced themselves as a possible final outcome of the meaning of what we do, and who we are as "rational beings." In short, whenever we attempt to speak with comprehension, with understanding, we accept the risk of not comprehending, of failing to understand; we even run the risk of losing our grip on what we had once understood so well, rendering the intelligible unintelligible—for to articulate, to put into words and thus to open the question of justification, is always to run the risk of our certainties becoming infected by uncertainty.

If so, then perhaps "crisis" should not be understood as an exceptional situation confronting science, but rather as the norm. Perhaps the point is to recognize that there has never been a situation in which the idea of science was not on some level an *issue*, a conviction always being pushed perilously close to doubt, where the obviousness of the idea always at the same time masks a failure to break through that very obviousness, justifying itself so as to ward off the specter of its own unintelligibility. Perhaps the history of reason in European culture shows us that rationality, however "naively" accepted it may have been and perhaps still is, has never failed to demand from us, at a later time or on another level, a decision or a willingness to struggle—including our own contemporary conflict with the tradition of reason, which for Husserl is nothing less than the struggle that has become definitive of European humanity:

> If man loses this faith [in reason], it means nothing less than the loss of faith "in himself," in his own true being. This true being is not something he always already has, with the self-evidence of the "I am," but something he only has and can have in the form of the struggle for his truth, the struggle to make himself true. True being is *everywhere* an ideal goal, a task of *episteme* or "reason," as opposed to being which through *doxa* is merely thought to be, unquestioned and "obvious."[39]

If this is the case, then a rigorous distinction between "crisis" and "critique" becomes less interesting, even misleading. To engage in crisis and critique is not somehow to respond in an exceptional way to an exceptional situation, e.g. to address an unfortunate necessity that has made itself felt because, somehow, the project of science has gotten off track. Instead, both crisis and critique are two moments within the normal situation of a life in truth, which is "everywhere" an engagement with the ideal of truth, understood as a task for which one has been made responsible. The task is understood fully only from the perspective of this responsibility, which implies a concern, a belief, an engagement; the ideal is "real" first and foremost as a possibility opened by such a responsibility, by having a stake in what the ideal "says," or better: what it promises. The ideal, from this point of view, is never a given actuality, if by "actual" is understood something fully resolved with respect to the question of what it "is"; instead, ideality is the function of a subject who looks beyond reality towards the possible, and who understands that it is precisely a "possibility" that is here at stake, or the issue at hand. "Crisis" could then be used as a term that would cover *both* the situation of conflict, of doubt and uncertainty over the question of the meaning of true being for those who speak on behalf of its realization, *and* the understanding that guides the one who is in the midst of just such a conflict—an understanding the "moderns" would normally designate as the function of critique.

However, these two senses of crisis are really not distinguishable, once we think of the conflict itself is the shape that has been assumed by an understanding of the world. Looked at this way, critical understanding is not something brought to bear on the crisis of the times, it *is* the crisis of our times, not merely a symptom but a manifestation of its very essence. For "critique" need be nothing more than simply another term for naming the participation in the crisis, in the questions that concern us, or which we face. And the exercise of critique, as an intellectual gesture, is simply another expression of being in a crisis, which is in turn for Husserl being concerned with the possibilities encountered in the wake of a commitment to the ideal of reason, including the possibility of its impossibility.

Above all, on the second reading it would make little sense to understand critique as a *cure* for the crisis of the sciences, as if there were such a thing as a special, philosophical therapy that would remove altogether the questionableness of science. Such a critique is, to be sure, a long way from being content with uncovering the unjustifiable limitations of objectivism; and moreover, it should be noted that Husserl never backed away from the assertion that the meaning of science in its "Wissenschaftlichkeit" can, in the end, be established once and for all, beyond question. To be sure, science as an infinite movement of world-discovery would remain as a task, but the meaning of this movement as a whole would be a firm acquisition arrived at by transcendental philosophy.[40] Such clarity is the ultimate goal of reflection as *Besinnung*: the sense of the task is what the philosopher seeks to re-establish, or to re-integrate into an understanding of what is at stake in the realization of the ideal.

Yet there is also a tendency in Husserl's writings of this period to suggest the opposite, that the very meaning of science, its function not only as the project of discovering the world but of opening the world to and for a "genuine" life in truth, is itself perpetually in a state of crisis. From the latter perspective, the opening of the world to understanding in the form of the true is never simply positive, but is always at the same time the opening of an experience of the questionableness of the world. For Husserl, the "whole" of all that is, as will be argued below, is presented originally by rational life in the form of a question. The manifestation of the world in its truth, thus in understanding, is first accomplished in the question "what is there?", and this accomplishment, with its originary uncertainty, is never left behind. It will also be argued below that this is what points to Husserl's insight into the necessity of an analysis of the concrete world of life. It is not simply because the empirical "fills out" the general, the ideal; rather, the empirical is a mode of positive being that, as a unique answer to the general opening/questionableness of the world, always preserves the question *as a question* within the meaning structures that determine the style of worldly givenness. To be sure, there are always answers to this or that question, but these are not answers that would rid world-life of its fundamental problematic character; the result being that there are only "answers" here in a special, limited sense. On this reading, if crisis is a disease, it is one for which there is no cure, critique or not; it would be like calling life itself a disease.

If this second reading is justified, then not only would there be little interest in drawing a distinction between "crisis" and "critique," but the unfolding logic of Husserl's reflections would themselves pose an inner source of resistance to such a distinction. A similar internal resistance can also be discerned in Aristotle's political philosophy as well. What for Aristotle would political life be, or better, what sense would it make to call *dikê*, a "virtue," if the manner of its exercise were to be decided in advance by critical theory, and presented to the citizen as something already resolved? At most, politics would be reduced to the simple application of a norm, the ultimate meaning of which has been and will always be decided in an "external" fashion—i.e., external to the political act. Then what would political action be, for a life in which there was no choice-saturated *krisis*—would this not, on the contrary, imply that such a life would have no need for political action, that virtue had been replaced by algorithmic mechanisms of decision without a hint of risk? Moreover, would one be able, for Aristotle, to speak of "citizens" at all, if there were no questions that demanded the exercise of *kritikê* within a horizon of uncertainty, where the task is to ultimately decide just what the sense, or meaning, of an action is going to be?[41] Would the man of practical wisdom be able to make the choices that a man of practical wisdom would make, if there were no practical choices to be found, just variables to plug into the latticework of equations constituting political method?

This second reading would also allow for a formulation of the motivation for a reflection on the meaning of science, in answer to the question posed above at the end of the Introduction. If crisis is not simply a singular event in time, but belongs to the structure of the realization of the idea of science in history, then this strongly suggests that the questionableness of science in turn belongs to its very sense. This is something I hope to show below in more detail, following the development of Husserl's text. If it is a viable interpretation, then one could argue that, for Husserl, what motivates reflection to ask what is possible, or what can be revealed about the possibility of science as such, is not something external to the meaning of science, but must be understood as a structural moment of this meaning itself. To engage the meaning of science would be to engage it in its very questionableness; to make the meaning of science clear, to articulate it and thus to expose oneself to its driving force—all of this entails a maturation of a sensitivity not only for the claim of science, but for the *problematic character* of this claim. This would mean that the sense of crisis is essential for the reflective engagement with the claim of science; it does not drive us away from the hold of the ideal, but on the contrary contains its secret. The first manifestation of this motivation is the crisis itself, which is not dispelled but rather *heightened* by *Besinnung*; the meaning of the ideal science, once it has been given a new force and presence thanks to phenomenological philosophy, will not a countermovement *against* the "crisis of the sciences," but its very culmination.

With this interpretation, one could also propose a possible reading of Husserl's description of the philosopher as a "functionary of mankind" in a difficult, and somewhat fragmentary, passage found in the section that closes Part I of the *Crisis*:

> We have become aware in the most general way [through the foregoing reflections] that human philosophizing and its results in the whole of man's existence mean anything but merely private or otherwise limited cultural goals. In *our* philosophizing, then—how can we avoid it?—we are *functionaries of mankind.* The quite personal responsibility of our own true being as philosophers, our inner personal vocation, bears within itself at the same time the responsibility for the true being of mankind; the latter is, necessarily, being toward the telos and can only come to realization, if at all, through philosophy—through us, if we are philosophers in all seriousness. Is there, in this existential "if," a way out? If not, what should we, who believe, do in order to be able to believe? We cannot seriously continue our previous philosophizing; it lets us hope only for philosophies, never for philosophy.[42]

Here both the readings outlined above seem to dovetail. The philosopher has a function, apparently, to administer to the developing cultural whole of humankind. Yet if we were to ask *how* such an administration would function, that is, how it is that the functionary would function, we come up against a tendency towards the second reading. For the question of the philosopher is not simply the question of humankind alone, but first and foremost of a *self,* of whoever reflects. It is from one's existence as a person that the role of functionary is possible at all; it is in an inner life that the question is first formed, and it is only in an inner life that it can be established, even if the responsibility entailed has a universal scope, and is not limited to the responsibility for oneself. And the question of the philosopher is above all that of the possibility of philosophy, which is something confronting the philosopher as a task of existence—though, to be sure, it is again a task that has implications beyond the individuality of the one who reflects, carrying with it significance for the telos of humanity as a whole. Alternatively, as Husserl puts it, this task will have significance only if we take it up, if we recognize the possibility of philosophy as that for which we ourselves are responsible.

If one were to choose to emphasize the tendency towards the second reading, then perhaps the term "citizen," in the Aristotelian sense of *politês*, would have been a better expression than "functionary."[43] The citizen is the one who is concerned, thus called to the task, which includes the possibility of failure, as well as risking the discovery of the meaninglessness or unintelligibility of the task as such. To be sure, "we philosophers," Husserl says, "believe" in reason—but that does not mean that we are not in crisis. On the contrary, as those who *believe*, it could be said that the crisis concerns "us" most of all. "Belief" is a particularly apt expression here, since it suggests that in the end the task remains, even for the confirmed citizen, an only partially fixed or real possibility, a decision that is only on the way to being decided. To believe is a genuine, but not a stable acquisition of life; it is to be actively engaged in the hold a meaning can exercise, but it is an engagement that in turn exposes one's self to all the indeterminacy, weaknesses, gaps, and inconsistencies that a meaning, experienced as a task, may bring.

Following this line of reflection, the last three sentences of the passage above can perhaps begin to become clearer. To be a philosopher is not simply to take on a task, through some act of will; one does not believe only thanks to the will. Questions of belief are more complicated. Thus if belief is descriptive of a response or attitude to the crisis, which itself could be described as a threat to bring a belief to an end, the point is not that this response amounts to an inner spiritual effort to remain loyal to an idea, to be "serious" about reason "to the end." The point is subtler. If understanding itself harbors a formative power, as I tried to highlight above with a consideration of Husserl's notion of the "surrounding world" (*Umwelt*), and if reflective engagement brings to the fore possibilities intrinsic to the hold of meaning, then a *Besinnung* with respect to the idea of science aims at the potential of a kind of spiritual rebirth, or re-creation of a future out of an exhausted belief. However, this re-birth is already inscribed in the belief itself, as part of its foundational structure; it makes us, who believe, able to believe. To be in a "crisis," then, is to be situated somewhere in between the capacity of our beliefs to make us believers and the historical possibility that this capacity, "for us," "in our case," may fail to determine our experience of the world.

This is, I would like to suggest, the answer to the question Husserl asks: "what should we, who *believe*, do in order to *be able* to believe?" To direct oneself to the question, what must be done to be able to believe, to be able to speak out of oneself on behalf of an ideal, is to approach the experience of understanding in its function of opening up a world, for this is the only context within which it is possible to "believe." We believe only in what we accomplish; to believe in science, to experience its force, can only be to realize a possible experience of the world. A decision for or against a belief conceived in isolation from this world-forming function of understanding, where such a belief becomes the object of an act of faith that takes place in some inner space of subjectivity, is out of the question. The point is instead to pursue a possibility within an unfolding, historical understanding of the world, within reason as that which has historically formed the world in which we live—above all perhaps within the cultural milieu of a *Gelehrtenrepublik*—and to engage understanding on *this* level, the level of the question of the origin of the world as such, is to engage in the very possibility of a belief in reason.

All of this is very far indeed from the spirit of the initial question with which Husserl begins his first installment of the *Crisis* for the Belgrade journal *Philosophia* in 1936—that is, the spirit of the scientist who, shrugging his shoulders in disbelief, asks: where is this crisis? For the crisis of the sciences is what science looks like, or how it presents itself, once one has learned to suspect the naiveté that characterizes such a question. Such naiveté is a kind of blindness, one that obscures what is actually involved or at stake; it is the kind of naiveté that would consider it obvious that a "citizen of Athens" is "someone who comes from Athens." To fully understand what is at stake in something is to grasp what is problematic about it, what is being risked and for what, thus what conflicts and decisions have defined it as a concrete life experience. To become cognizant of the task of the crisis demands the clarity won by the shedding of naiveté—the naiveté that would be content with the belief that I am where I am from, or that I believe what I believe, and that is all there is to it, and that there is no particular problem involved in being who I am.

"Naiveté" is not meant here to be a term of reproach. There is in fact enormous power in naiveté, and it should be respected. Furthermore, it is arguably the most important concept of Part II of the *Crisis*, and its thematization eventually leads to the possibility of transcendental phenomenology itself. Thus in this sense it is necessary to return to the question Husserl poses in the beginning about whether there is truly a crisis at all; or at least return to it by reflecting on the significance of the fact that the crisis can be something hidden, latent—despite the fact that it is always present as a structural moment of the meaning of science, as I would like to argue is the case.

In answer to the question how a reflection on the meaning of science as such could be motivated, the second reading proposed above, which will guide the rest of this study, suggests that reflection itself belongs to a natural progression of an awareness of the crisis of reason, of the growing necessity of asking the question whether that in which we believe is in fact possible at all. However, it will be the concept of naiveté, of the obviousness of things, including the ideal of science, that will provide Husserl with the insight as to how it is that science as such, even structured as an historical crisis, functions as a formative understanding of the world. Only once the essential role of naiveté is fixed will it be possible to fully make sense of Husserl's promise of a rebirth of the ideal of reason in phenomenological philosophy.

Notes

[1] For more on the Cercle, Husserl's trip to Prague, and the role of J. Patočka in the visit, see Jan Patočka, *Texte—Dokumente—Bibliographie* (1999) 176-257.

[2] Hua XXIX 103-139 (Text Nr. 10, "Die Psychologie in der Krise der Europäischen Wissenschaft. <Die Präger Vorträge>).

[3] Crisis 3; Hua VI 1:4-6. Cf. Hua XXIX 103:6-13.

[4] Hua XXIX 103:8-10.

[5] Crisis 4; Hua VI 1:22-23.

[6] In the texts of the Crisis period, Husserl does not always lay as much emphasis on this aspect of success as I am doing here. Often the argument will be that the crisis of science is due to the failure of science to take into account the lifeworld. Cf. P. Janssen, *Grundlagen der wissenschaftlichen Welterkenntnis* (1977) §33, for an emphasis on this line of argument in Husserl's texts. For a nuanced reading that comes closer to the one presented here, see R. Thurnher, "Husserl's Idee einer 'wirklichen' echten Wissenschaftstheorie," in: *Krise der Wissenschaften—Wissenschaft der Krisis?* (1998) 37-38.

[7] Crisis 6; Hua VI 4:5-17.

[8] Hua XXIX 150:16-20.

[9] As was E. Fink, Husserl's assistant and close collaborator during this period. Take for example the summary at the end of the working manuscript of the Prague lectures, which the editor of Hua XXIX indicates was marked by Fink as "Entwurf von Fink," and where there appears the following lines: "Für den kühnen und verwegenen Versuch des neuzeitlichen Menschen, die Wissenschaft zur entscheidenden Lebensmacht seines Daseins werden zu lassen, ist in unserem Zeitalter die Schicksalsstunde gekommen. Es wird sich im Laufe e i n e s Menschenalters erweisen müssen, ob die Lebensgrundlagen des Europäers sich endgültig verlagert haben, ob er wirklich sein eigenes Wesen preisgibt oder ob die

allgemeine Hinwendung zu irrationalen Mächten nur Symptome einer tiefen Erschöpfung sind—was also die Hoffnung in sich schlösse, daß durch eine Radikalisierung der Selbstbesinnung eine Neubegründung einer wahrhaft autonomen, d.i. philosophischen Kultur noch möglich sein könnte." Hua XXIX 137:36-138:11. It is tempting to see Heidegger's influence in these lines, though that would not be fair to either Husserl or Fink. It should simply be recognized that, politically, both Husserl and Fink were conservative German nationalists of a particular kind, and that they were at their philosophical best where they left their political sentiments out of the picture.

[10] The title of the lecture is "Die Philosophie in der Krisis der europäischen Menschheit," (Hua VI 314ff) which, in somewhat free form, was delivered by Husserl on 7 May 1935, and repeated 10 May due to popular demand. See "Einleitung des Herausgebers," Hua VI xiii-xiv.

[11] Crisis 272-273; Hua VI 317:38-318:8.

[12] On the general conception of regional ontology, see Ideas I/Hua III,1 §9; also cf. Ideas II/Hua IV §§49-53, on the distinction (influenced by Windelband, Rickert, Dilthey and others) between ontologies of "nature" and "spirit."

[13] This is the guiding thesis of Husserl's *Experience and Judgment*: knowledge as a body of logical formations is not something that comes into play only once a subject actively thinks, or makes judgments, but is always already in play in a multiplicity of different ways before thinking takes place as a series of explicit subjective or cognitive "acts." To be sure, the description of the "logical" achievements of pre-predicative experience requires a different conceptual vocabulary than is the case with logic proper; they are not themselves explicitly cognitive achievements, but achievements of sense that opens the field of possibility for cognition itself. Cf. EJ/EU §§5-11.

[14] One that is, of course, transcendentally intersubjective: "Sie [Welt] ist konstitutives Gebilde der Intersubjektivität, in Unterstufe der vergemeinschaftet erfahrenden, vorstellenden, evtl. bewährenden, evtl. sogar theoretische Wahrheit zwecktätig anstrebenden Subjektivität [...]." Hua XXIX 258:8-11.

[15] Although the ambiguity does not necessarily disappear on the transcendental level. Here the thesis is re-asserted: consciousness never simply finds its objects, but always becomes "conscious of" things in the mode of an achievement (*Leistung*, *Sinngebung*). Nevertheless, the world is not *produced* by consciousness in the literal sense, even if its manifestation is nevertheless a transcendental achievement of consciousness. Cf. Ideas I/Hua III,1 §§55, 149; also cf. Crisis/Hua VI §§48-50.

[16] This extends to the possibility of being responsible for this sense, to again refer to a passage quoted in the Introduction: "Die höchste Stufe der Rationalität als bewußter Selbstrationalisierung und, was gleichkommt, als bewußter Erbauer seiner Umwelt (und der Idee nach ist das schließlich die Kulturwelt überhaupt) zu einer Kalokagathie, die er ihr gegeben hat als die rational zu verantwortende, ist die aus einer Philosophie, und zwar einer Philosophie, die nicht ein Sport von Einzelperson, ein Bedürfnis ihrer Individualität ist, ihrer Wißbegier, ihres Bildungsinteresses, sondern einer Philosophie, in der der einzelne Träger der Menschheitsbesinnung ist im Dienst der höchst zu verantwortenden, der absolut zu wollenden, gewollten Selbst- und Weltgestaltung." Hua XXIX 226: 11-25.

[17] Another expression for "the world in the mode of being what it is" is "the world in-itself" (*Ansich*) or "being in-itself" (*Ansichsein*). Cf. Hua XI §§23-25. In these lectures, Husserl presents the passive, pre-givenness of the world in subjective receptivity as something that is already the space of an original productivity, *prior* to the entrance of the "I" as the subject which turns towards things in the mode of cognitive interest.

[18] That point being made, the role of "picture" in this experience of world-embodiment that is science should not be underestimated. Cf. Heidegger, "Die Zeit des Weltbildes" (1938), in: *Holzwege* (1950).

[19] Perhaps one might object: Does not Husserl himself call science a "garb of ideas" (*Ideenkleid*)? He does (see Crisis 51; Hua VI 51:29), but the question of just in what sense science is a distortion of the world of experience is hardly settled. The point is not that concepts distort life instead of explaining it; that is a relatively trivial issue. The real question is the place and function of such distortions—this is what is to be accounted for in the *Crisis* through a reflection on the concreteness of life. Concrete life will be the point of departure for the explanation of its own distortion.

[20] Not that we cannot play games. It is just that Husserl rejects that it could be satisfying philosophically to think of thinking as a game. See FTL/Hua XVII §33.

[21] Crisis 6; Hua VI 4:27-29.

[22] See Text Nr. 25, "<Die> Welt der Tatsachen (9. und 10. August 1936)," Hua XXIX 293-301.

[23] Crisis 6-7; Hua VI 4:29-5:1.

[24] And is it not, on some level, basically Nietzschean? That is, where science is an example of a valuation, an ordering of the world, that *devalues itself*, negating its own effort to be that which provides life with the world in which it can exist. This point of contact is all the more interesting when Husserl, in an equally profound *anti*-Nietzschean turn of phrase, claims that the task of these reflections is to turn back to an originary self-understanding which, as "ultimate, original, and genuine," exercises a hold on us that, once "seen" in a clarfied vision, "apodictically conquers the will." (Crisis 18; Hua VI 16:18-19.) No will to power here.

[25] The tenor of the scene is set by the mockery of human wisdom by Mephistopheles, along with Faust's own doubts, which he expresses at line 1810: "Ich fühl's, vergebens hab ich alle Schätze / Des Menschengeists auf mich herbeigerafft, / Und wenn ich mich am Ende niedersetze, / Quillt innerlich doch keine neue Kraft; / Ich bin nicht um ein Haar breit höher, / Bin dem Unendlichen nicht näher."

[26] That is, Husserl will describe science, and philosophy as determined by an inner teleology. Cf. Crisis 70-73, 299; Hua VI 71:28-73:23, 347:13-32; and Hua XXIX 362-420 (Text Nr. 32, "Teleologie in der Philosophiegeschichte").

[27] Or is it just as "far" from the ideal? A close reading of the "Origin of Geometry" could suggest the latter. Husserl is open to the possibility that the task of re-establishing a living connection to the ideal of science, thus to its inwardly motivating telos, is attainable by recourse to genetic questions not because practitioners of science in the past were closer to this ideal, but because the genetic-historical mode of questioning itself is conducive from our perspective towards attaining the goal of originality. After all, if it is due to the distortions of historical life that we are separated from the ideal, that is something that only we are in a position to comprehend, not those who stood at the primal "beginning" of science, whose sense of the history of science could take the form only of an *anticipation*. This issue will be the central concern of Chapter Four, below.

[28] The normative priority of the possible with respect to the question of legitimacy or validity, and its intimate bond with the movement of actualization, will be an important theme below. Cf. A. de Muralt, *L'Idée de la phénoménologie* (1958) 55: "C'est pourquoi Husserl peut dire en fin de compte que ce qui légitime le fail actuel de la science es sa subsomption sous l'idée de la science possible. Par là il indique manifestement que *c'est le possible qui norme le fait (actuel)*, alors que nous venons de définir l'idée-norme comme acte."

[29] See R. Koselleck, "Krise," in: *Historisches Wörterbuch der Philosophie.*
[30] A point of comparison with Augustine in this use of the term can be made with respect to the structure of the "situation" of crisis for both authors. For Augustine, the crisis of faith is not the crisis of a radical choice between belief or non-belief, as if it were in one's power to simply "choose" to believe. The crisis is a decisive moment that occurs within a belief that is already given—the choice of acceptance or non-acceptance occurs in the context of grace, within which the ability to accept has already been given to man by God. And if for Husserl the "crisis" is a "choice" between rationality and irrationality, it is still a choice made by a rational being—a being who already is that which is to be either rejected or accepted. Cf. Augustine, *Confessions*, ch. VI.
[31] *Politês d' aplôs audeni tôn allôn orizetai mallon he tô metechein kriseôs kai archês*.See also *Politics* 1253a and *Nicomachean Ethics* 1134a31.
[32] Also of importance, particularly since a key part of Husserl's argument in the *Crisis* is an interpretation of the development of European philosophy from the Renaissance, is Koselleck's *Kritik und Krise* (1959).
[33] For Schiller, the realm commanded by such a critic is not so much politics as the stage, where the sovereignty of the critic does not detract from, but instead aids that of the prince: "Eine merkwürdige Klasse von Menschen hat Ursache, dankbarer als alle übrigen gegen die Bühne zu seyn. Hier nur hören die Großen der Welt, was sie nie oder selten hören—Wahrheit; was sie nie oder selten sehen, sehen sie hier—den Menschen." Friedrich Schiller, "Was kann eine gute stehende Schaubühne eigentlich wirken?" in: *Schiller's Werke.* Bd. 20 (1962) 97. Cf. also Koselleck's discussion of Schiller's 1784 lecture in his *Kritik und Krise* (1959) 84-86.
[34] Cf. *Nicomachean Ethics* 1134a31: *hê gar dikê krisis tou dikaiou kai tou adikou*; "justice is the decision/ conflict/ choice/ determination of the just and the unjust."
[35] Crisis 12-13; Hua VI 10:35-11:10. Which already raises questions. In other writings, Husserl makes the point that skepticism is not simply false, but in fact comes closest to the "critical" perspective ultimately represented by phenomenology itself. Not that phenomenology is skepticism, but because the historical significance of skepticism represents an initial, if flawed rejection of *naivete*. Cf. Hua VIII/1 8ff. Also see the discussion by A. Aguirre in his *Genetische Phänomenologie und Reduktion* (1970) 65-116.
[36] See especially Hua VI 315:11-19.
[37] Crisis 12; Hua VI 10:3-21.
[38] Crisis 13; Hua VI 11:23-30.
[39] Crisis 13; Hua VI 11:11-18.
[40] See in particular Crisis 335-341 (Appendix IV); Hua VI §73.
[41] Here we can cite the second of Jan Patočka's *Heretical Essays*, "The Beginning of History," in particular the rich description, inspired by Heraclitus, of the common origin of philosophy and the *polis*. Political action, in the freedom of an exposed humanity that is no longer confined in the natural rhythms of the family, is itself a grasp of the nature of the whole in its ur-manifestation as *polemos*: strife, conflict. But so is seeing into the nature of things, or philosophy itself— "Thus *phronesis*, understanding, by the very nature of things, cannot but be at once common and conflicted." *Heretical Essays in the Philosophy of History* (1996) 42.
[42] Crisis 17; Hua VI 15:24-16:3.
[43] There is a passage from a text composed by Husserl in the 1920's where the term "Funktionär" is used to designate membership in a community organized around a common task: see Hua VI 300:24-301:2.

Chapter Two

The Manifold Sense of Foundation (*Crisis* §15)

Establishment and sedimentation

Part II of the *Crisis* (§§8-27) is concerned with the historical transformation of the idea of philosophy that takes place in the course of the development of modern science since the 17th century. The theme of Part I has not been left behind, as long as the concept of crisis is understood to be a contribution to the description of the historical character of the idea of reason. Moreover, it could be argued that it is in fact the positive character of the crisis that is the real theme in Part II, and which constitutes perhaps the most difficult insight pursued by Husserl in these paragraphs. The difficulty is in part the result of a certain ambiguity about the "idea of science" when considered from the point of view of the question of how an idea can be, or even should be an object of historical reflections. The question is fundamental: in what sense is an "idea" present to consciousness thanks to its history? In what way does an idea owe its manifestation to history, such that it makes sense to approach it in an historical reflection, as opposed to a purely conceptual reflection?

Husserl's position will be radical, arguing in effect that an idea is genuinely present only in history, or in its historical becoming. For Husserl, the idea of science does not, when taken in itself, in its "ideal" independence from the movement of its own historical realization, exhibit a perfect coherence of sense. This is because the idea is dependent on its expression in order to be, not what it is, but manifest "for us"; we genuinely touch the idea only in a perfection that is manifest, however indirectly and incompletely, in a gesture, or an expression. An expression is something the actual realization of which is guided by an approach to an idea. As the guide of an expression, the idea is in this sense "present." However, this presence of the idea is only progressively realized in the expression, the reality of the idea *is* the very progression of the expression towards its idea, which means that the idea takes on the characteristic of being "genuine" only in the unfolding of its historical expression. Perfection is manifest in the progressive being-perfected of an expression. The consequence is that it is only in the being-perfected of a gesture or expression, in which the idea has taken a form that renders it visible, that the idea can be said to be understood; in this way, the "history of the idea" is the history of its "being understood," and in this sense the history of its appearance in the ongoing perfection of an expression of the understanding. But this also means that the "realization" of the idea as such never finds its resting place in a concrete, achieved perfection; from the very beginning, as was already argued in Part I, the idea begins to collapse, its very being-perfected or rendered visible in the perfections of expression is accompanied by a growing unintelligibility. By no means does this imply that perfection is thereby excluded from playing a role in understanding the meaning of science, nor does it mean that perfection has no existence; the point is only that there is never a clean, unambiguous presence of the idea, where it would stand as its own unambiguous exemplar.[1]

A related ambiguity is explored by Husserl in his *Formal and Transcendental Logic* (1929). A purely formal logic, even when it has been successfully articulated in the form of a pure formal ontology of the "something in general" (*Etwas überhaupt*), cannot ultimately satisfy the interest in a perfected understanding of the world; for the perfection of formal analytics is ultimately "ideal," without being at the same time "real." The systematics of pure logic understood as formal ontology remain wholly within the space of the pure formally ideal being-understood of the given; but the scientist is ultimately interested in actual truth, in the truth of "given" being. Truth in its actuality, in being the truth of the given, is not simply the filling out of what is possible, as if all that there could be said "genuinely" of the given were already accomplished in a discourse of that which, at best, only potentially is, of the ideal "something in general." In this way, the perfection of pure analytics accomplished by modern axiomatics, if taken to be the realization of the ideal of science, can only threaten to obscure the issue by substituting itself for the genuine perfection of a genuine understanding of the world of things. In the turn from the analytic truths of the pure formal constructions of the "anything whatever" to the actually experienced, given truth of beings, to "what is" in its becoming, there is a turn to a dimension of the understanding, and with that to a dimension of ideality itself, that cannot be identified as "mathematical" without serious distortion.[2] It is precisely the orientation within scientific consciousness to this "given truth," and with that to an understanding that grasps the truth of the "given" world in light of the idea of science, which is the principal theme of Part II of the *Crisis*.

All of this of course raises many questions. What I propose to do here is to outline some basic interpretative concepts to guide a reading of Part II, taking as a point of orientation §15 of the *Crisis*, which bears the title "Reflection (*Reflexion*) on the method of our historical manner of investigation." Here Husserl pauses in his presentation and takes a step back, in order to clarify what he is looking for in his discussion of Galileo and early modern science, which makes up §§8-14.[3]

A prominent concept in §15, as well as in many manuscripts from this period, is that of *Stiftung*. The term *Stiftung* is often translated using variations of the English word "establishment." *Stiftung* can also simply mean "foundation"—as in a financial basis for the support of an activity, or an endowment for a university or field of research. Greek science, for Husserl, can be said to represent a *Stiftung*, insofar as it is the store of wisdom upon which the entire edifice of European thought and culture is established or founded. However, Husserl's use of the term can be rather complex, as is the case in §15, where *Stiftung* is being used as a metaphor for history. The movement of European history itself is being described, and not European culture as a collection of ideas and values that form a coherent "picture" of the world, and that would be subject to movement only in its factual construction and preservation. Greek science is a founding, to be sure, and even a foundation, but these need not be taken in the sense of a gathered basis, or a static reservoir of wisdom in which basic principles have been secured. The temptation to reconfigure the metaphor in architectural terms should be resisted. When one speaks of the "reason for" or "purpose of" a project, for example, one means something very different than when one speaks of the "foundation" of a building; the former is pliant and open, the latter rigid and set. The reason for a project is its original motivation, the grasp of purpose used to chart out a particular course of development, and which can be refreshed at any subsequent phase; whereas the foundation of a building is forever what it first is, an immobility that functions as the key support for an entire structure. It is the former sense that is operative in §15: Greek philosophy is the "foundation" for a project, in that it represents the original motivation that projects a certain course of understanding. Greek science founds, to use other language, the *task* of science; it does not merely support the *results* of science by securing its premises.

Husserl thus uses the language of a "task," in particular a task that has been "assigned" by way of an originary establishment. To say that a task has been assigned for Husserl is in fact just another way to express the sense in which it has been "founded," "instituted," or "established." But what does it mean, to assign a task? It is not simply to fix or to give "something to do"—instead, it is, to invoke another important expression often used by Husserl in the *Crisis*, to call (*rufen*) on someone to act. If the task is fundamental, then it is a call to exist in a certain fashion, to be concerned with a question, an issue (in the sense of "crisis" explored above). And if one were to argue that understanding and reason itself should be understood as a task, then what is meant by a "task" is not the performance of this or that activity alone, but also the state of having-been-assigned the task itself. This aspect can be emphasized by thinking of a task as an activity that has been given weight thanks to a call for it to be performed. As assigned, the task is in part the expression of this call, which, giving weight to the assignment, is the origin of a kind of *necessity*.

This necessity sets activity in motion, it "necessitates" it; which perhaps expresses a more fundamental sense of the necessary than simply the fact that something "must be" or "has to be." To assign a task, or "establish" it historically, is to imbue it with a necessity, understood as a kind of impetus, a drive that sets into motion possibilities that need not already be decided, fixed and determined in advance. Possibility, in other words, can be in motion in the assignment before it has settled into a fixed definition, which also means that an establishment can be open to renewal. This also provides another expression for the existential themes that play such a conspicuous role in Part I of the *Crisis*: before it is decided what we must be, even before the possibility of being has been fixed or determined, necessity, thanks to the assignment of a task, has already made it possible for us to be at all; in an important sense it makes possible before any sense in which it is decided what, in being so necessitated, it is thereby necessary to be. When Husserl inquires into the possibility of the task of philosophy, I would argue, he has both of these senses of necessity in mind, and in the order of this distinction: it is first a question of a calling, a vocation, that makes philosophical life possible at all, setting it into motion as a possibility of existence; then, and only then, is it a question of being compelled by what is necessarily implied in the very calling of the task, as it gradually becomes clear in its unfolding just what it is to be.

Nevertheless, the assignment of a task alone is nothing historical. To be sure, to be assigned a task is possible only for a being who is "in time," for a task is something which at least in part lies in a "future," it can even be considered as one of the forms in which the future itself is present, determining yet at the same time not determining its content. Yet, by itself this does not mean that the task that has been assigned is a specifically *historical* task. That would suggest that its ultimate fulfillment is something to which an individual, or the "we" of a generation, merely contribute, that it is something the reality of which is sustained by resources that transcend the singular, limited event of one's life. Historical tasks are shouldered only by historical persons. Thus something more is needed in order to bring the theme of history to bear; for it is not the case that the mere sense of having been assigned a task can only be made evident if we reflect on ourselves as beings who exist not only in time, but in history as well. That would again seem to imply that the task is not first and foremost directed, assigned, or made possible "for us," but for an historical subjectivity that would be more what we are than we ourselves, or the self to which we must turn in order to understand why we are being called on to do what we are being called on to do.

Thus, history truly first comes into play only once our relation to that which has been assigned to us has been experienced as a peculiar kind of problem. In fact, with respect to the task of science, this relation, according to Husserl, is not a simple, unmediated relation of being addressed. The task that "necessitates" us is not simply, directly "addressed" to us in a literal sense, even by a call that has been preserved in the cultural memory that has recorded the Greek insistence on a life in reason, which again and again exhorts each new generation to fulfill the task of living a rational life. It is not, in other words, one choice among others, that may be taken up or not; we cannot remain indifferent. This is because the call, for Husserl, is on the *inside*, insofar as what we are, or who we already are, is an essential part of understanding what is *meant* by the call, or how it is that we stand in relation to the possibility of pursuing such a task at all. On the one hand, why it is *we* who are being called is not a mystery; we stand in this relation because we are already a history, where this history is indeed more what we are than are ourselves, a life that belongs to our very core even if it is something that we do not always manage to fully come to terms with. On the other hand, in order to come to terms with this history, and thus ourselves, an *inner critique* is required in order to form a genuine relation to our own historical existence, to the task that has been assigned to us; and it will in turn provide insight into the role that this history plays in bringing the task to bear on life. Looked at in this way, history does not come from outside, pulling us into its flow and assigning to us a purpose that is not originally our own but forced upon us by the chances of past transmissions, but emerges as an inner unity of sense that we discover as our own, on the "personal" level:

> In a constant critique, which always regards the total historical complex as a personal one, we are attempting ulti-

> mately to discern the historical task which we can acknowledge as the only one which is personally our own. This we seek to discern not from the outside, from facts, as if the temporal becoming in which we ourselves have evolved were merely an external causal series. Rather, we seek to discern it from the *inside*. Only in this way can we, who not only have a spiritual heritage but have become what we are thoroughly and exclusively in a historical-spiritual manner, have a task which is truly our own. We obtain it not through the critique of some present or handed-down system, of some scientific or pre-scientific Weltanschauung (which might as well be Chinese, in the end), but only through a critical understanding of the total unity of history—*our* history.[4]

The unity of the task, evident only as the guiding idea of an unfolding unity of a history, underscores the simultaneity but difference of the second sense of necessity, i.e. the necessity that determines and not simply sets into motion. It is this necessity of the task in the second sense which we also, gradually and from the inside, come to understand as the true, "ultimate" sense of being-assigned a task. For the clarity of what it is that calls us, and which we arrive at by an inner critique, is in the end the clarity of the task itself, what it had always been seeking as its rightful and not simply contingent place in our lives:

> For it [our history] has spiritual unity through the unity and driving force of the task which, in the historical process—in the thinking of those who philosophize for one another and with one another across time—seeks to move through the various stages of obscurity toward satisfying clarity until it finally works its way through to perfect insight. Then the task stands before us not merely as factually required but as a task assigned to us, the present-day philosophers.[5]

Why this is so, why our fundamental relation to our ownmost task is historical in character, or only becomes "ours" by way of a clarification of the unity of historical being ("teleology," "idea"), is not addressed in §15, and is in general only partially answered in the *Crisis* as a whole. The question still remains: does this idea of an "inner history" commit Husserl to the position that there is no level of subjective life, of subjective being, that is not exposed to the historical, and with that compromised in its freedom, in the very integrity of its personality—or in other words, is there not a real sense in which the tasks assigned to us by our past are in fact, in the end, "Chinese"? At most one can say that there is an important tension between, on the one hand, Husserl's development of the theme of history and, on the other, his unrelenting focus on the personal dimension of philosophical life. For Husserl, it is always persons who are the catalysts, but what they realize is always more than their own; this is perhaps true of *Besinnung* most of all.

What can be found in §15, however, is an important discussion of the consequences of the historical character of this relation to the possibility of being called, in particular a fundamental tendency towards *obscurity* that marks any manifestation of the task as a task. The theme of obscurity and distortion is something I would like to link to the sketch above of the thesis of the irremovability of the element of crisis that I argued can be discerned in Part I: any historical understanding or sense for the task, any experience of being held in the "sway" of the inner teleological unity of history (its very "idea"), involves a dimension of crisis.

To describe the systemic generation of this obscurity, Husserl employs another metaphor, this time a geological one: "sedimentation" (*Sedimentierung*).[6] As with the case of *Stiftung*, this metaphor can be as misleading as it can be helpful. What is meant is not that history accumulates strata as layers on a cake, where the 17th century would be sandwiched somewhere in between the Greeks and us. Nor is the metaphor meant to express what one might call the narrative structure of history, where the "story" of the past, in which one epoch follows another in a linear fashion, can in turn described as the origins of "deposits" or strata of the ideas, sensibilities, and consciousness that make up the existing surrounding world as if it were some great archive. In such a view, the historian, as the one who tells the story, could be likened to an archaeologist excavating the sense-depths of the cultural world that we inhabit as historical beings. Husserl uses the metaphor of sedimentation in quite a different sense, which can best be emphasized by avoiding the image of the world as a *structure*, an object that has such and such a constitution in place, ready to be analyzed (or excavated). It is better to think of the world as the earth, or as that ever present though obscure ground beneath our feet, which in everyday life we neither tend nor need to make into an object at all, but which is nevertheless "there" as an *obscure mass*.

My point in invoking this image of "earth" as an obscure mass is to emphasize a non-architectonic dimension of the metaphor of sediment, which is important for many of Husserl's texts in the period of the *Crisis*, even if he does not always stress it. Sedimentation begins, one could say, when something is being packed down by something that has been put on top of it, thus pressed into a layer or sediment of a larger mass which, as a whole of such layers or sediments, remains a whole by virtue not of an order, but of a force that presses these layers together. Now, "history" is a kind of mass, it could be said, in an extended sense; it is not simply a string of events, but events that are somehow "packed together," kept in place by the weight or burden of historical life. This description could be used to distinguish the historical past from the past as such, where the former could be said to *bear down* on the present in a way the latter does not. As such a unity, history can thus be said to be a whole of parts that fit together, not necessarily because they belong together or have been constructed narratively, but because they have been pressed together by the momentum of history itself. History in this way has a double character, it both moves and stands still, and it is both a burden and a release. What, in the case of history, are the elements, the "events" that are being pressed into a unity? And what state are they in, once they have been pressed? Moreover, what does it mean, for this "earth" to move?

Let me make a suggestion, while again stressing that this is meant as a preliminary reflection towards developing an interpretive strategy of Husserl's text. The suggestion is this: it is not the events themselves that become sedimented, or pressed together; rather, it is the understanding of the world that has been shaped by events, the manner in which we understand and encounter things in the wake of events that take place, not in the world, but in the inner life of understanding. One aspect of the process in which this awareness takes shape is sedimentation. What is sedimented together is what we have come to "know" about the world; what we know, and how we know, settles in place in the wake of a developing history of knowledge. The result is that this knowledge is never simply "in place," but in being put into place, "sedimented," it has taken on a particular character, or undergoes a formation: it has become a *concrete* pattern of the surrounding world.

One way to describe this is to say that sedimentation begins when what we know about the world, i.e. what we know because of our experiences with a particular way of knowing and thinking, takes on the character of what is *familiar* to us. This brings us to what is perhaps the single most important concept of the *Crisis*: "familiarity" or "obviousness" (*Selbstverständlichkeit*). The familiar is the obscure mass of a sedimented history. When something becomes familiar, it becomes pressed into, or fused with the familiarity that is the whole, the obscure mass of our surrounding world of earthly existence. And here we have "earth" used as a metaphor not in the sense of an object of an historical archaeology, but the earth in the sense of the solidity, density, and mass of the ground beneath our feet, that ubiquitous horizon of what is familiar, or *taken for granted.*

Sedimentation thus describes a kind of establishment, but not in the sense of *Stiftung*, which is above all an act that has set us in a certain direction. What is established in sedimentation is nothing assigned, nothing that could be present as a task or a vocation; if anything is being "fixed" here, it is that something is being "taken for..."—taken for granted, taken as a given in a passive intentionality of conscious life. This passivity is not secondary but originary; the sedimented is a given in the sense of something to which we need not pay any attention, as if it has already been decided long ago, a decision that need not be made in order to be. This is what Husserl calls *pre-given* (*vorgegeben*): the pre-given is a given that never broaches on the questionable, thus which remains within the circle of what is familiar, as its center of gravity.

One has the tendency to become familiar in this sense with whatever the understanding has succeeded in articulating; it is the success of a project of thinking that leads to this particular kind of obscurity. This also includes whatever may result from the fact of the understanding having been assigned a task, thus from its activity in the wake of a *Stiftung*, or in response to its call. Even the understanding of the task itself, i.e. the task of fixing the sense of what it is to understand, is subject to this being "pre-given," or given in the mode of the familiar. Thus *Stiftung* and *Sedimentierung* are two aspects of the same process of the formation of the surrounding world: once I discover a calling, I become familiar with the patterns of behavior and thought that follow from its idea; for example, once I become a writer, I eventually learn to "live the life of a writer," so naturally that I no longer have to think about what it means to write. At some point—"historical"—I simply am what it is that I have become; the idea that guides my becoming has become "familiar," a pattern I do not even need to think about in order to manifest, or be the manifestation of. And if one describes this obviousness as a kind of "obscurity," such an obscurity does not necessarily involve an obliviousness, as if in feeling at home in the life that is the result of a calling would entail in any sense a forgetting. On the contrary: I continue to deal with that with which I am now familiar; only now it has become a non-problematic part of my surrounding world, nothing about which I ever find cause to *think.*

This obscurity of the familiar entails a certain freedom. When already familiar with something, I do not need to bring it continuously into view; it remains there, working in the background as something I already know "all about," and thus have taken to be too "obvious" to warrant any attention. It is an understanding that, in having fulfilled itself, no longer requires that it be exercised. There is thus here a kind of cessation of understanding, where nevertheless the effect of understanding remains; for not all such cessation need be understood as a forgetting, or a lack of understanding.

However, obviousness is not simply negative, in the sense of the absence, resulting from a success, of the need for explicit attempts to understand. It is also productive or formative of the activity of understanding as such. Once something is familiar, thus by implication no longer strange, or a question that remains to be addressed, the result is that I am in a position to see implications that would have remained invisible were the obvious not obvious. Familiarity is solid ground on which to stand, it is something on which to rely; and it is reliable for an historical existence precisely as sedimented ground, as a pressed ground of static understanding that allows us to look beyond it to possibilities that we can only now, standing on solid ground, take up and realize.

With these metaphors of establishment and sedimentation, Husserl pursues an analysis of the manner in which a concrete mass of the world of understanding is a condition for the possibility of the realization of an idea, that of reason itself. It is the very "obviousness" of the world that sets into place a horizon in which understanding can be pursued as a task of reason, where reason can establish itself as a calling. Moreover, it is an "inner critique" of this obviousness, of that on which we rely as a basis of unthought understanding, which will articulate our historical relation to the content of this call.

Why is critique, and with that *Besinnung*, necessary? We have here the outlines of a more sophisticated version of the basic concept of crisis. For it is obviousness itself, and its concreteness, that poses a risk of what Husserl calls *Sinnesveräusserlichung*,[7] or a separation from the reliability not of the sedimented mass of understanding, but of the meanings that have become embodied therein. That is, the realization of new possibilities of understanding the world, themselves "grounded" on obviousness or pregivenness, can set the stage for the experience of breaks between the sense of the world that has become obvious and the sense of the world which has been newly acquired, or understood thanks to the former. These breaks are not faults in a line of deductions, but result from a lack of sensitivity for what has been accomplished, for the connections between the obvious and what is accomplished in the unfolding project of understanding. For in the end, the obviousness on which we rely in order to see further also hides things, like Poe's purloined letter safely concealed in being openly displayed. Like the letter, it is not that the things we have come to understand are forgotten or lost; they are hidden within the very visibility they have thanks to reason—within the very visibility of the world around us, the world we inhabit. This invisibility in visibility, or our loss of comprehension in a world we have come to fully understand, is the "crisis of European sciences." Thus "externalization of sense" occurs wholly within the inner space of the understanding, it is a crisis of the inner life, where the clarity that has been achieved is not necessarily a clarity that reveals things, but which stands in a complex relationship to the sense of things as things. Here again is the paradox I indicated in Chapter One, but given a potentially more fruitful description: the very familiarity that opens for us the world at the same time introduces the danger of a strange obscurity. It is strange, for it is not an obscurity that stands opposed to clarity, which the understanding, as a power, could attempt to penetrate; rather, it is a systemic obscurity within the historical unfolding of clarity as such.

In light of this, the meaning of "historical reflection" for Husserl is essentially twofold. First, it is the thematization of the forward progression of scientific understanding, which Husserl describes as "teleological," insofar as what is assigned as a task is defined by the idea of science. Nevertheless, history, as a movement that realizes the idea, also hides it, since as a progression of familiarization it continually works to hide the unity of what has been accomplished in the clarity of our understanding from the awareness of those who understand. That is, it hides, so to speak, the meaning of the accomplishments of the understanding in the very obviousness and familiarity of those same accomplishments. Thus the task of these reflections is in part to recognize the accomplishments that science could make only insofar as it was historical; but the task here is also to defeat history, or to make visible what history, as that mass of obscure familiarity that shapes our experience of the surrounding world, necessarily hides. Ultimately, for Husserl, the telos of the history of reason—the history guided by the idea of reason, which is also the history in which this idea is manifest—is towards a knowing, an understanding, that sets itself up beyond history, and which is in that sense its culmination.

This reflection, in revealing the manner in which we are related to the idea of science as a task ("historical"), in turn sets the stage for the next form of the call itself, or what is to be demanded from us, what is "necessary." That is, it is this *defeat of history* that we are called on to perform; this is what is required by the idea or telos that defines our historical being as such. To fix our historical relation to the task of science through inner critique is to clarify historical life of all naiveté, and in this sense to free life from a certain kind of historical experience. And it is *this* task, assigned by a *Stiftung* made in full awareness of the obscurity of history and acting against it, that gives birth to what Husserl calls "modern philosophical humanity."

Ur-Stiftung, Nach-Stiftung, End-Stiftung

With these observations and suggestions, we are now ready to consider two passages from §15 in which Husserl sketches the structure of the assignment of this task, when considered as an historical phenomenon. In the first passage, the term "functionary" makes another appearance, but this time in conjunction with the theme of *Stiftung*:

> For we are what we are as functionaries of modern philosophical humanity; we are heirs and cobearers of the direction of the will which pervades this humanity; we have become this through a primal establishment [*Urstiftung*] which is at once a reestablishment [*Nachstiftung*] and a modification of the Greek primal establishment. In the latter lies the teleological beginning, the true birth of the European spirit as such.[8]
>
> [...]
> But to every primal establishment [*Urstiftung*] essentially belongs a final establishment [*Endstiftung*] assigned as a task to the historical process. This final establishment is accomplished when the task is brought to consummate clarity and thus to an apodeictic method which, in every step of achievement, is a constant avenue to new steps having the character of absolute success, i.e., the character of apodeictic steps. At this point philosophy, as an infinite task, would have arrived at its apodeictic beginning, its horizon of apodeictic forward movement.[9]

It is important to be clear that the problem here is nothing less than the beginning of philosophy. Not the "beginning" in the sense of the problem of identifying the first historical philosophers, but in the sense of a need to ask how in general it is possible that philosophy has a beginning, how its sense can be something that has been arrived at, or definitively "put into place." To be sure, it is not at all clear why such a question is important. Could not this question simply be rejected as meaningless, that the task of philosophy has simply been given to us by the tradition that has, quite naturally, already to a great extent decided the meaning of philosophy? To some extent, such a claim would be justified. The tradition has in fact already given us the project or task of philosophy; not necessarily in the form of a set of rules or directions, but nevertheless as a reasonably well defined course of thinking, characterized by a concern with true being, with the question of the good, and so on. Yet, Husserl's claim is that to be in possession of a tradition means to be engaged with the problems that have to do with what this tradition means *for us*—more, the way we relate to the tradition, the place it has in our lives and our culture, is something historically variable.

Thus, the tradition itself is to be understood as a kind of task, one that we can approach or receive only given a more basic decision about the meaning of history. This includes that aspect of historical existence that is so familiar to us as to be invisible—familiarity itself does not mean that all of the questions have been answered, but it does mean that some questions have to be discovered, or perhaps re-discovered as questions. It is this historicity that lies at the root of the beginning as a problem, or which turns the assignment of the task into the problem of a beginning. The beginning, already present, or already having "happened," nevertheless has to be established, made clear, or else the full import of the task itself will be hidden in the obviousness of familiarity, like the excitement of a story being lost by its being told too many times. In a paradoxical sense, philosophy attempts to somehow begin before its own beginning, to ask questions before it is clear that such questions can be asked at all; this is the age-old philosophical practice of inquiring into the obvious, or insisting on the weight of a question even where its answer is fully evident. This is no less true of the "obviousness" of philosophy itself: philosophy must first come to an understanding of its own possibility, of its own beginning, which means that it must face the possibility of its own impossibility. A kind of skepticism is thus built into the very fabric of philosophy, the result of which is that there can be no "traditional" philosophy.

The emergence of the problem of the beginning of philosophy itself takes part in the founding, the *Stiftung*, of the historical task assigned to scientific humanity. "Historical reflection," in which the question of the beginning is formed, is not a "mere" reflection on the history of an idea, but a re-articulation of the compelling quality of the idea, thus a rediscovery of the task assigned by the idea. We thus remain within the purview of reflection/critique as *Besinnung*. In this way, historical reflection, having taken the form of critique, must itself be recognized as belonging to the movement of history, though of an "inner" history. In the passages I have just quoted, Husserl charts the trajectory of this movement by employing three different senses of "foundation" (*Stiftung*): (1) *original founding* (*Ur-stiftung*), (2) *re-founding* (*Nach-stiftung*), and (3) *final found-*

ing (*End-stiftung*). They can be understood as three different aspects of the beginning of philosophy as it is refracted through the lens of historical reflection.

The original founding (*Urstiftung*) is a "first," and thus genuine beginning, but in a peculiar sense. The genuine beginning in which the idea of philosophy is experienced is not a self-conscious beginning; it is not aware of itself *as* a beginning. The very theme of *Urstiftung* is meaningful only within an orientation to the problem of the beginning, or the search for the beginning—thus precisely where the beginning has been called into question, or is in need of being questioned. The "originary" character of this beginning as an historical problem is thus valid "for us" alone; it does not apply to the Greeks, for they did not have any sense of being a beginning *for us*—the Greeks were not aware of the sense in which they would someday become *our* Greeks.[10] This does not exclude the self-understanding of the Greeks as unique, or even as the "first" to accomplish advances in scientific thinking (all of which is more difficult to establish than it may seem). Nor is what is meant the mere historical truism that those who come before can never be familiar with those who come later, that we can never really know the impact we will have on the future. The point turns on the sense in which the appropriation of history itself bears on our relation to the idea of science, of reason as such; there is more at stake than a familiarity with the past. "Original foundation" is a genuine philosophical theme only for those who are conscious of the necessity of approaching the question of the beginning as a historical problem, i.e., precisely for those who have become aware that the relation to the idea and task of science is an historical relation.

The same holds true for the "re-founding" (*Nachstiftung*), which is in itself an original founding, but one played out in a broader context. The re-founding is a "new beginning," precisely in the sense of a re-naissance. It has the unique, and paradoxical feature of realizing something that has already been, but as if it were new. It holds up as a model possibilities realized by the original beginning, and projects further possibilities that, on the foundation of former achievements, have only now become clear as possibilities. Still, re-birth is still a birth; it breaks new ground, and is not simply a repetition or appropriation; it is the same, born anew, in a different world; it employs and values the same purposes of the original birth, but now "made over" into new purposes for a new time and context. As a birth, a re-birth is "original," thus we can and must speak of the "Nach-stiftung" as itself an "Ur-stiftung."[11] Furthermore, as will be seen when we turn to the *Origin of Geometry*, it is also the case that every original foundation must also be considered a rebirth, in the sense that *anything* historically new plays off whatever had been "in place" before. Again, this points to the importance of the metaphor of sedimentation: every sense of the original is relative to something that, as a prior originariness, has become sedimented into the familiarity of the "surrounding world."

Yet for Husserl, the "birth" and "re-birth" of the task, its founding and re-founding are visible only from within a reflection on the possibility of a finality, of a clear sense of what it is that undergoes this pattern of historical emergence and re-emergence. It should again be stressed that it is only when the founding, the *Stiftung*, has become a problem of the beginning, and is not taken to be a positive "given" pure and simple, that it can appear in this threefold sense—which is really not composed of three separate senses, but rather what one might call a historically sensitive anticipation of the possibility of a "genuine" beginning. The formulation of the problem is historically sensitive insofar as the possibility of philosophy is taken to be not only realized, but also anticipated, even to the extent to which the historical realization of philosophy is recognized as really only the historical realization of an anticipation—an outlook towards a finality, or a culmination of the idea, the task. And if the finality of the *End-stiftung* is fixed in the form of its own apodeictic character, then the very sense of this apodeicity can be said to have been formed historically, or within the concretion of a relation to the ideal that is possible only in the form of a history.

If so, then "historical being," "historical existence," is not blind, nor does it pose an obstacle that must be overcome in order to discover the apodeictic character of science. On the contrary, the very sense of finality, of the "End-", rests upon the intentionality that takes the form of continual birth and re-birth, or precisely the form of "history." Thus if, as it was put above, we must also defeat history, it is only because the history of the idea of science contains in its very sense a reference to the task of its own defeat; which implies that to "understand" this history is precisely to understand the necessity of overcoming this history. Even if the "beginning" of philosophy is ultimately an historical phenomenon, and its most genuine historical manifestation takes the form of an historical reflection, it is nevertheless the beginning of a decisive break with the historical.

The finality arrived at in the historically conscious emergence of the beginning of philosophy is of course a function of phenomenology itself. But it is important to pose the question of the possibility of phenomenology, of its "beginning," with caution. The point is not to turn to history and discover what had to have been already in place, or "thought," in order for phenomenology to be possible. Rather, the question is to instead come to an understanding of the concrete sense of the possibility of phenomenological philosophy for "us," for "European civilization," as a living tradition in which this intention or anticipation of the idea of reason has historically taken shape. To arrive at an understanding of this sense, of what it means for phenomenology to be possible (or, in general, for science itself to be possible), requires a reflection that articulates what is left unsaid in the naïve, "natural" sense of what it is to pursue science. It is here, in the sedimentations of understanding, that one has a feel for the idea that makes itself available as a beginning; but it is only an explicit question, an issue that must be addressed, because of the crisis itself. It is the crisis, the "sense-externalization" where the idea of reason appears as alien within its very manifestation, that has put philosophy and science as a whole into relief as a problem motivating the reflection on historical sense, precisely in order to discover in what sense one can speak any longer of the *demonstrability*, and not just the *familiarity*, of this possibility of reason for "us," for our lives.

To be sure, this raises questions. If, for example, the apodeicity of phenomenology, as the consummation or "final form" of the scientific attitude, is in some sense at least a break with the obscurities of historical sedimentation, then how is it that a reflection on history is going to reveal the possibility of such a break? Does history, historical being, really anticipate its own transcendence; does it really occur in constant anticipation of its own being superceded? How is history, as we put it above, both a burden *and* a release? If the very *sense* of this break, thus "finality," is intimately bound to that from which it breaks, then in what sense can we speak of a "break" with history at all? Or will the break itself remain somehow unaccountable, providing the subject with a position over the movement of the historical that reveals its relativity, but all the while immune to the consequences of its own genesis in this relativity itself?[12]

These questions will be pursued throughout the remainder of this study, though they will necessarily take a different form as our reading of the *Crisis* develops. For now we have from §15 a structure that we can use to orient ourselves with respect not only to Part II, but Part IIIA of the *Crisis* as well: the threefold sense of founding as *Urstiftung*, *Nachstiftung*, and *Endstiftung*—originary founding, re-founding, and final founding. Chapter Three will begin where Husserl himself begins Part II of the *Crisis*: Galileo, who represents for Husserl the re-birth of the task of Greek science, thus a case in point of the *Nachstiftung* of the idea of science that is definitive of modernity. We will then turn in Chapter Four to the question of "origin" (*Ur-sprung*) itself which, for Husserl, is always present along with any re-founding. Finally, in Chapters Five and Six we will turn to the question of the perspective of phenomenology itself as the formulation of the problem of the "final form" of scientific philosophy.

Notes

[1] Cf. A. de Muralt, *L'Idée de la phénoménologie* (1958) §§3-14. In this study the question of teleology in Husserl, which is perhaps the most intractable issue in the *Crisis*, will not be presented in all of its aspects. The attempt here is rather to lay the groundwork for understanding in what context it makes sense, in a phenomenological reflection, to speak of a "teleology of reason." To a great extent, the same can be said for the *Crisis* itself. To begin with the concept of teleology in order to understand the unity of the history of reason would be misleading; it would also undermine a grasp of the inner relation between crisis and telos for Husserl.

[2] This is the basic outline of an argument that recurs throughout *Formal and Transcendental Logic*. One important example is how Husserl uses it to underwrite the retention of a philosophical conception of logic as apophantics, even where apophantics is structurally/formally equivalent to formal ontology. Husserl argues that apophantics ultimately retains a groundedness in the original motivation of and for logic as a whole, which is precisely an interest in the truth of the given or concrete; to collapse apophantics into formal ontology would be a further obfuscation of the origin of the *ultimate sense* of logic as such, even if from a technical point of view there would be no difference. See FTL/Hua XVII §44.

[3] It should be noted that §§16-27 will not be dealt with here in any detail. These sections treat the philosophical expression given to the insights and resulting ambiguities of Galilean science in Descartes, Berkelely, Hume, and Kant. One reason for this is that these sections are sketchy at best. What is original in Part II of the *Crisis* is the discussion of Galileo, but more importantly the conception of historical reflection from a phenomenological point of view. Nevertheless, the theses presented in §§16-27 are not empty speculations, but represent an outline of Husserl's reflections on the history of philosophy over the previous decade; cf. in particular the lectures from 1923-1924 (Hua VII).

[4] Crisis 70-71; Hua VI 72:2-17.

[5] Crisis 71; Hua VI 72:17-24.

[6] This metaphor is employed by Husserl in a number of contexts (for example at Ideas II/Hua IV §§29, 56a, Beilage XII §1; EJ/EU §§25, 33, 50b, 67, 81a; Crisis/Hua VI §9h, Appendix VI / Beilage III), but is perhaps developed the most in a manuscript from 1931-1933 where Husserl describes the constitution of the unconscious as the establishment of the "surrounding night of the sedimented" ("Umgebung der Nacht des Sedimentierten"). D14, 4a.

[7] Crisis 44; Hua VI 43:22-23.

[8] Crisis 71; Hua VI 72:24-30.

[9] Crisis 72; Hua VI 73:32-74:1.

[10] Cf. Hua XXIX 45-51, where Husserl defends the role of "Dichtung" in the history of philosophy—especially Hua XXIX 50:35-51:2: "[Es ist das, <was> Wirksamkeit übt:] im 'Verkehr' mit den Phantasiemenschen erfahren wir Wirkungen, die oft tiefer eingreifend sind als die der für uns wirklichen Menschen." The Plato we construct in our historical imagination, Husserl argues, is a far more influential personage than the actual, "historical" Plato who remains invisible to us. Though the question should be posed: for "our" Greeks,

with all of their sensitivity to the poetic, the rhetorical, to manifestation and appearance generally—is it really so certain that they could be so oblivious to becoming "our" Greeks?
[11] Cf. in particular Hua XXIX 421-423 (Text Nr. 33, "Die Unterscheidung zwischen absoluter und relativer Urstiftung").
[12] P. Ricoeur asks a similar question: "Does an Idea or a task in general develop a true history? Does an advent make an event? The paradox of the notion of history is that it becomes incomprehensible if it is not a unique history unified by a sense; yet it loses its very historicity if it is not an unforeseeable adventure. One way, there would be no more philosophy of history; the other way, no more history." *Husserl. An Analysis of his Phenomenology* (1967) 170.

Chapter Three

Galileo and Modern Science (*Crisis* §§8-10)

The history of "obviousness" (*Selbstverständlichkeit*)

The thesis formulated at the end of the last chapter is that, for Husserl, the final form of philosophy, "phenomenology," can be fully understood only in terms of its historical character, and only realized through an historical reflection. Its meaning, and thus possibility gains clarity only given the turn to historical reflections of the type represented by *Besinnung*. But what do we turn to, when we turn to the history of an idea? The question is all the more pressing if one accepts the notion of an idea emerging in history, even that it only emerges thanks to history; this excludes the possibility of a fully formed idea somehow being encountered outside of an experience of an historical reality. To turn to the history of the idea, for Husserl, amounts in the end to simply turning to the idea itself. How is this to be understood?

If the idea, at any given phase of its history, is only an incompletely manifest sense, then perhaps the history of an idea can be understood in terms of the multiplicity of the modes of its *anticipation*. Then it would be a reflection on these anticipations that would be necessary for the sense of the idea to become clear. But if so, if finality of form, or "clarity," can be accomplished only given an awareness that engages this historical dimension, then it would be as if clarity itself had its ultimate, grounding origin in the constitution of an anticipatory intentionality. Again there is something paradoxical here: The idea is fully manifest in its clarity—thus genuinely "is" for us—only when we are sensitive to the manner in which it has been anticipated in its unclarity.

Once "anticipation" has been employed in historical description (along with establishment and sedimentation), one should also be sensitive to the quite legitimate point that whatever is anticipated can be truly understood only from the perspective of the fulfilled, realized phenomenon, which is precisely no longer present as something anticipated. An anticipation is intrinsically provisional, thus it cannot stand alone as a secure establishment of clarity. Only against the ideal limit of the positive presence of the actual does what "came before" become visible at all, even as an anticipation. This means that to suggest that the meaning of the idea in the present is clarified only with a turn to history amounts to putting oneself in a circle, since the historical present, as the fulfillment of the past, is *also* the clarified future of that past. The turn to history as a mode of clarification operates only in the present, which means only once it has itself been clarified by its own "outcome." As a result, historical reflection cannot be a stable perspective, operating independently from the present; it is not a "view" on the present from outside the present, one that could compete with others, challenging their truth. For one thing, from the perspective of the historical past itself, fixed according to its inner logic, the "future" or outcome that is the present is nothing that could be said to have determinate shape; the historical past does not represent a way to "know" the present, even in the form of that which "will come." The present is, in a peculiar sense, *not in view* when it comes to history; an historical narrative that disrupts this ignorance characteristic of the past with respect to the present immediately loses its credibility. The historical present is, again, at best anticipated by the past, not strictly determined or fixed; which means that it can clarify the present only when, in a basic sense, its own inner logic is superseded by that of the present.

Despite its paradoxical character, the theme of anticipation is essential for historical description, because it can be used to highlight a sense in which, through a reflection on the bond of present and past as that of fulfillment and anticipation, the present can be said to be *revealed to itself.* It is precisely this self-revelation of the present through its own history that represents the pattern of clarification that Husserl is after in Part II of the *Crisis*. It represents a pattern founded in implications inherent to the very sense of "having" a history. If the "historical meaning" of the present is at least in part constituted by what can be discerned as the anticipation of itself in the historical past, then this opens the way to the insight that the status of the present, as a "result" or "outcome," is not autonomous, but dependent upon its having a past that, in its "own time," in turn "had" the historical present projected before itself in the form of an anticipation. The perspective of the past is not thereby granted an autonomy that it itself does not have; rather, the center of gravity within the perspective of the present shifts, in such a way that the historicity of the present itself becomes a new ground of visibility, precisely in and for the present. Having a past, and being aware of itself as something anticipated by the past, the present becomes uniquely visible to itself. The insight thus made possible is the recognition of a dependency of the sense of what is "firmly" in hand (the historical present) on a prior state of indeterminacy (the anticipations of the past), where the very possibility of what is historically present is (or "was") being formed. Not only is the historical meaning of the present thereby visible, but its historicity as well, as the ground of its meaning.

Something else goes with this. To recall one of the themes of Chapter One, one of the tasks of historical reflection is to grasp the historical indeterminacy of the present in its being projected by the past in light of its *productive capacity*, that is, the capacity to shape the surrounding world. What has come before does not fashion the present (that is up to the present), but it does fashion the possibility of the present, above all of the inner sense in which the present "has" possibilities. This is the genuine goal of the historical-teleological reflections in the *Crisis*: it is not to explain the present, to demonstrate the origins of what now exists, but to reveal what the present *could be* in light of the history to which its potential can be traced as its origin. Thus, historical reflection in Husserl's sense is radically different from historiography, or the science of history, which, one could say, seeks to make real the past by securing knowledge of the past as an empirical component of the shared cultural world. Instead, the point is to make real not the history of philosophy, but the possibility of present philosophy through its history; once this possibility is in hand, the present will itself be re-opened, re-articulated, and above all re-projected in the form of the task of a philosophical life.

How is this done, how is the anticipation of the present by its own historical past isolated for reflection? And once isolated, how does such an anticipation "clarify" a possibility, in the radical sense of revealing the "productive" origin of that same possibility? That is, how is the historical anticipation of the idea of science made a theme, in a such a way that what is thereby made a theme takes the form of an open possibility, which at the same time has weight and significance "for us"?

At first glance, it would appear that Husserl in Part II simply adopts the stance of an historian of ideas, or the standpoint from which the relation between the present and the past is understood in terms of the influence of the one over the other. The present idea of science is one variant among many in a long history of the development of "concepts of reality" and "methods of research." To be sure, the "history" in Part II is clearly conceptual history, and anticipations are described as similarities that arise between past systems of philosophy and present approaches, where the latter represent unanticipated advances over or developments of aspects of the former. Nevertheless, Husserl introduces a key innovation by orienting the entire question of anticipation around the concept of obviousness (*Selbstverständlichkeit*), which I have already highlighted in the preliminary discussions of the last chapter. Husserl transposes the usual investigations of the forerunners of modern philosophy (their theories, techniques, discoveries, failures) to the question of the origin of what is "obvious" to us. In particular, the central question of Part II has to do with the familiar fact that nature, the world as it is understood rationally, is "mathematical." The point is not to tell the story of the first historical persons who came up with this idea; rather, the point is to address the question "what is the meaning of this mathematization of nature"[1] by reflecting on the origin of the obviousness of this meaning for us.[2] This introduces an important element to the description of anticipations: the issue is not to find forerunners who anticipated the idea of the mathematization of nature, but to fix the inner historical anticipations, the thought-forms and tendencies of an intellectual world in which it is normal, or "obvious," to think about nature in this way.

Above it was suggested that the obvious, or that which is taken for granted, plays a productive role in the understanding of the world, insofar as what is taken for granted acts as a foundation for thought. The ground upon which Galileo stood, what Galileo took for granted, opened up for him a field of possibility within which he pursued his science—and it is precisely the axis of "ground" and "possibility," or what Galileo took for granted and what he saw to be possible, that is the theme for that reflection which seeks to formulate the anticipation of the future (the historical present). The idea is that what was obvious in the past was (in part) the point of origin for what is obvious for us; the "ground" of the familiar is not something fixed or stable, but developing, where the movement of this development represents the fashioning of possibilities within the invisible passivity of the obvious.

But how is this ground that was, this obviousness of what Galileo took for granted, accessible at all, i.e., to those who reflect historically? Husserl's strategy in these pages is in fact a tenuous one. For the basis of his reflections cannot be the reports and documents handed down from the past, the usual "basis" for guiding, and above all legitimating historical inquiry. For no one documents what is obvious, so in this sense archives are all but useless; to take something for granted is to be indifferent to any concern or call for preservation. A history of obviousness, then, is a uniquely invisible history. Husserl's point of departure in such a reflection can only be what is *for us* obvious; the obvious cleaves so close to the contours of experience that it is impossible to discern its forms outside of the immediate self-presence of experience as such. This implies that, paradoxically, the obviousness that is our own is the only connection that we have with what was the ground of obviousness for Galileo. The result is that the history of obviousness is a kind of pure historical phenomenon, since it has no extra-historical manifestation that would assure us of its existence—there are no manuscripts or monuments to the obvious, there is only an inner, historical-intentional interpenetration of "our" obviousness and that of Galileo. But if so, if what was obvious to Galileo has not been preserved or recorded independently of our own prejudices, then why does that mean that it is not simply lost? Is it not in the end a sophism to look to what is obvious to us, in order to investigate what was obvious to Galileo?

Everything depends on the claim that there is a genuine connection between what is obvious to us and what was obvious to Galileo. And such a claim, if in many ways uncertain, is by no means unreasonable. For the historical present is, after all, the consummation of the past; it is a world that has been fashioned within an historical space that was once marked out by the anticipations of past humanity, and is inhabited by those who are now trying to clarify the possibilities of this now actual world by posing the problem of a reflection on the anticipations of the past. More specifically: what is obvious today, science as the science of mathematical nature, *became* obvious through an historical movement or tendency that had its origin in "Galileo," in the sense that Galileo, acting within the horizon of possibility of "his" world, formulated or anticipated the horizon in which the possibility of our own historical present has taken shape. Husserl's argument is that a key parameter circumscribing this movement is the ground upon which Galileo stood, what he took for granted. However unspoken, undocumented, implicit and invisible, what was obvious for Galileo has left its mark on what he was able to accomplish because of it, and thus cannot be relegated outside of the dynamics of the historicity of the present without distorting its intentional structure.

The connection is a genuine, perhaps even the most genuine connection to the past, but as a basis for reflection it involves obvious difficulties. First is the problem of circularity that has already been mentioned. If the genuine historical sense of what is presently "obvious" to us is to be articulated, and not merely by a narrative undermined by anachronisms, then we must proceed in such a manner that we maintain the sense of *difference* between what we have come to take for granted and what was taken for granted by Galileo:

> We should devote a careful exposition to what was involved in this "obviousness" for Galileo and to whatever else was taken for granted by him in order to motivate the

> idea of a mathematical knowledge of nature in his new sense. We note that he, the philosopher of nature and the "trail-blazer" of physics, was not yet a physicist in the full present-day sense; that his thinking did not, like that of our mathematicians and mathematical physicists, move in the sphere of symbolism, far removed from intuition; and that we must not attribute to him what, through him and the further historical development, has become "obvious" to us.[3]

In order to capture the historical sense of what is obvious to us, we must take care to keep in view this difference between Galileo's situation and our own. But before we take on the difficult question of how this can be done, how we can truthfully say that we are able to draw this distinction at all in a manner that is useful to the problem at hand (the problem of clarifying our own possibility in terms of a past, and perhaps invisible anticipation), we need to heed what would appear to be a contradictory demand. That is, at the same time that we seek to keep apart Galileo's situation from our own, we must nevertheless continuously refer to the latter in order to make sense of the former. We find ourselves, again, "in a sort of circle":

> The understanding of the beginnings is to be gained fully only by starting with science as given in present-day form, looking back at its development. But in the absence of an understanding of the *beginnings* the development is mute as a *development of meaning*. Thus we have no other choice than to proceed forward and backward in a zigzag pattern; the one must help the other in an interplay. Relative clarification on the one side brings some elucidation on the other, which in turn casts light back on the former.[4]

Thus it is necessary to keep the Galilean historical situation separate from our own, being careful not to treat what is "obvious" for us as if it were "obvious" for Galileo; but at the same time, what Galileo made possible, thus an aspect of what he was, cannot be understood apart from our own situation, or what we are. We must keep Galileo separate from us, precisely in order to recognize that he is a part of us.

And it will be the concept of "obviousness" itself that will enable Husserl to do this. This is because obviousness is a kind of consciousness as well as a kind of unconsciousness, and in playing the one off the other,

Husserl will situate his reflections in the interplay, as he puts it, between the past and the present.

First, let us consider more closely obviousness as a mode of consciousness, or awareness. What is obvious, what is "taken for granted," can be likened to the confines of the surrounding world, in the sense of the whole of that which is *accepted* as the world. Whatever has been given a place within the sphere of awareness is "there" in accordance with the way that things are expected to be—we expect the world to have a certain manner, events to unfold in a certain rhythm of familiarity. But the unfolding of the familiar world is not set into a static pattern. Even if we leave aside the case of the unexpected, or what does not yet have a place, it is clear that the world is not necessarily confined by our acceptance of it, in the sense were acceptance would involve a kind of static definition. Acceptance itself changes; familiarity itself develops without breaking the fold of obviousness. The sense unity of the familiar world continuously develops beyond the confines of its own accepted familiarity; as a result, the "world" is more the open horizon of its own confinement, always somehow more than the whole, which it nevertheless is in its familiarity.

Historical consciousness is in part the recognition of this relativity of the confinement of the world within a certain style, or character of being accepted; it is the sense that what is taken for granted or familiar is not immobile, but subject to shifts and transformations that give us a feeling of the passage of time, or rather: of history. This relativity is present in the concrete sense that we have of a given thing as characterized by its own openness: there is always a level, even if superficial, where the meaning of the thing can shift, leaving behind what it once meant and taking on a new look, or aspect. There is no sense determination of a thing that is not open to either being replaced outright or at least modified, given a different look. The same is true of the world as a whole. Thus obviousness, as a species of "awareness," is the natural state of a finite consciousness that is potentially on the move towards something new, another re-confinement or revaluation of whatever new constellation of the obvious has shown itself to be possible—more, to be *given*. And if we are to be able to catch this germination of a new possibility in the old, then we need to be able to isolate what "had been" the relative confinement of the world, or how it was that the world was accepted "as world," precisely in order to discern its own movement away from itself, thus towards what it itself "makes possible."

On the other hand, obviousness is also an unconsciousness, or an unawareness of just what the limits of this peculiar confinement of the world are. That the world is confined in the mode of the obvious also means that its confinement is implicit, not explicit—I never follow these confines with my eye, mapping out a definite shape of what circumscribes all that is worldly. Or, at least, whatever is made explicit in the attempt to theoretically grasp once and for all "the world as it is," is accomplished in an activity that stands on the ground of an understanding that is more an implicit acceptance of the world "as it is" than an explicit formulation of what has been accepted to be the case. If so, then the task of historical reflection is not simply to isolate the relativity of the world-confinement expressed in Galilean science, but to bring out what is implicit, unconscious in this obviousness. Thus the point is not simply to reflect on how Galileo must have seen the world; the point is rather to bring out something that was *not seen* by Galileo, that *did not need to be seen* precisely because it was "taken for granted," accepted in the absence of any motivation to see.

"Obviousness," as an awareness of the world, implies an awareness of the historicity of the obvious; and as an implicit acceptance, it poses the task of explicitly elaborating what has been accepted. Historical reflection in Husserl's sense is thus the attempt to represent and analyze the logic of a *movement of acceptation*, a movement that is already in our grasp precisely in the form of what is "obvious" to us, taken in its full spectrum. We have the obvious, what is obvious for us, in our grasp as a movement; which means that we are already in possession of an awareness of what separates the situation of Galileo from our own, namely the natural movements and developments of obviousness itself. But what of the task of seeing what Galileo did not see? What in our situation provides us with the motivation to see not only what Galileo in his situation had not been motivated to see, but which, in our situation, may not even be there to see, apart from the mark it has left on the historical presence of Galileo?

Perhaps we are already in a position to answer this question. Based upon the discussion of §15 in the last chapter, one could say that it is precisely this need to see, to see what Galileo had no need to see, that begins with Galileo but only comes to fruition in "our" time, our philosophical situation of crisis. What Galilean science made possible, or anticipated, is precisely a modern science that can no longer simply have as its task the understanding of the world. It is now faced with the growing sense that the task of knowledge must also include the understanding of the meaning of knowledge itself, the "epistemological problem" in its modern form. The "beginning" of this need to see (which need not be interpreted as the "first" beginning) lies in Galilean science, in the concrete attitude towards the confines of the world that is its natural movement. Yet this beginning is nothing that is part of the "world view" of Galileo, but lies precisely within a dimension of Galilean scientific discovery and articulation of the world that is not, for Galileo, "in view" at all. The beginning of the future in his present is as invisible to Galileo as the future itself, as has already been stressed above. This beginning is present as a motivating factor only once Galilean science becomes historical—that is, only once it becomes visible as an anticipation of the present, a visibility through which the present reveals itself. In other words, the beginning must be brought out, in critical fashion, by those in the historical present who reflect; they are motivated to look for it, precisely because of the radical need to see what is implicit in what is taken for granted today, what has been accepted in the wake of the movement of history. We are motivated to see what Galileo did not see, precisely by the need to take no longer anything for granted, and to understand what such a need entails, or the possibilities for which it speaks.

What Galileo took for granted

Husserl outlines three things that Galilean science takes for granted; or, better, three accomplishments that form the accepted confines of the Galilean world. They are: (1) pure geometry; (2) the art of measurement; (3) the style of the world.[5]

(1) What is taken for granted about pure geometry? Husserl is not arguing that Galileo's acceptance of pure geometry as a legitimate science is irrational, or unfounded; on the contrary. The point is rather that, once the particulars of geometry as a science are more or less worked out, they take on the character of a readily available "body of knowledge" to which the scientist can return again and again as "the same"—which simply means that geometry is a positive historical acquisition. But that also means that its availability as already acquired, in place, does not need to be identified with the original availability of geometry, or the availability in principle of the science of geometry as a possible way to understand and articulate the nature of space, thus a possibility for thinking that presents itself not as an accomplishment, but as a *project*. On the contrary, once established, the availability of pure geometry is no longer that of a pure possibility for thinking, but is rather there for the scientist as an actuality. It is there as a path thinking has already traversed, a set of propositions already substantiated, where for the most part all of the "results" have been arrived at, constituting at least the principal theoretical basis for a fully articulated science. And these results are as available as anything else that is already an acknowledged part of the world; they are there "on hand" as recognized elements of its presence. Galileo, in short, was born into an intellectual world for which there was already a "geometry."

This difference between geometrical science as a project, a pure possibility for theoretical thinking, and geometrical science as an acquisition may seem trivial, but for Husserl it is of decisive importance. This is because the availability of pure geometry as an acquisition opens up the meaning of geometry, or the geometrical, to sense-developments that go beyond what is contained in the pure sense of its theories and principles taken in isolation. The availability of geometry as a cultural acquisition is not limited to the activity of geometrical thinking as such, for "acquired" geometry can also be used for other things. It can take on instrumental value, perhaps for other types of theoretical activity; it can act as a model, or a guide for thinking in general; it can have practical application; or it can serve as an insight into a sense of the world that had hitherto been obscure. Once geometry is in place, the horizons of the geometrical can open up beyond pure geometry itself.

(2) Also part of the surrounding cultural world for Galileo is another mathematical acquisition, but of a different sort, more primordial than that of geometry: the art of measurement. Pure geometry is founded upon a mode of idealization that fixes pure shapes as entities in and for themselves, ideal objects of a generality that is independent from what can be experienced; it is thus possible only given a direction of focus, an interest, that "abstracts" from the concretely real, positing (methodically, beginning with definitions, postulates, and common notions) the plenum of geometrical formations. The art of measurement, which is also mathematical and with that a mode of idealization, is motivated by a different interest, that of fixing the shapes of the experienced world "for us," enabling us to arrive at a common understanding of what is in fact the case. Measurement fixes the shapes of the world in terms of idealizations that articulate "how" the world is, or more specifically, in terms of "how much"—how much distance, how much time, how much weight. In its being fixed in this way, the world becomes for the first time "our" world, an intersubjective world wherein we can now speak of what is "in fact." In measurement, facts themselves for the first time become properly "objective facts."[6]

The world itself, then, in the very shape that it has for us in intersubjective experience, is already a mathematical world, insofar as it is measured and measurable—and this is in turn already an accepted character of the world. To occur within the confines of this world is to be measurable; what is immeasurable is unfathomable.[7] And if this is "obvious," always already accepted, then is it not also obvious to look to another mathematical accomplishment, that of acquired geometry itself, as a guide for the development of more sophisticated techniques of measurement, thus to use geometry as a tool for practical dealings with the world? It is, or at least it can be. But this further use of an already given geometry as an accomplishment need not be limited to the practical (for example, calculating the distance traversed by a projectile if fired from a certain angle), but could perhaps even play a more decisive role in the theoretical. For after all, geometry is the result of the successful pursuit of a theoretical interest, the interest in the systematic comprehension of geometrical objects. What about the interest in the comprehension of the things we encounter in experience, "physical" things? Is not that also a theoretical, and not simply a practical interest? It is, or at least it could be seen as such. The option, then, or the concrete possibility of treating the geometrization of the arts of measurement as the successful acquisition of a knowledge of the world the style of which is akin to that of the knowledge of geometrical objects—thus knowledge of the world *more geometrico*—is just as "obvious," or at least it could be.

(3) There is a third element Husserl emphasizes in his exposition, however, one that is both elusive and decisive. There is obviously more to the world than shape and form, and more to the sense of the given than can be directly captured in a naïve description that employs the mathematics of spatial structures. Geometrical being is abstract being, thus to approach nature as if its essence were geometrical is to argue that it must be approached as if it were abstract. And like anything initially expressed in an abstract way, the implications of what is really being "said" in such an expression are left, to a great extent, undecided, and remain so until a certain connection with the concrete can be re-established. It is here that Galilean science is truly unique in the history of ideas: the problem it pursues is not how to coordinate a Platonic mathematical realm with the non-mathematical physical world using a sophisticated application of hypothetical methods (as was the project of Ptolemaic astronomy), but to rediscover the concrete in light of its deep, hidden connection to mathematical structures, leading to the conviction that nature itself, in its concreteness and not just in its being-thought, is intrinsically mathematical. In Galilean science, therefore, there is no connection being made between abstract geometrical form and concrete quantities ascertained by the art of measurement, but between abstract geometrical form and something far more elusive and general, though nonetheless concrete: the style of the world as such. For Galilean science, not only is the essence of the world mathematical, but above all, it *behaves* mathematically.

This notion of the "style of the world" is intrinsically connected to the theme of "surrounding world" or *Umwelt* discussed in Chapter One, as can be seen from the following passage:

> The things of the intuited surrounding world (always taken as they are intuitively there for us in everyday life and

> count as actual) have, so to speak, their "habits"—they behave similarly under typically similar circumstances. If we take the intuitable world as a whole, in the flowing present in which it is straightforwardly there for us, it has even as a whole its "habit," i.e., that of continuing habitually as it has up to now. Thus our empirically intuited surrounding world has an empirical over-all style. However we may change the world in imagination or represent to ourselves the future course of the world, unknown to us, in terms of its possibilities, "as it might be," we necessarily represent it according to the style in which we have, and up to now have had, the world. We can become explicitly conscious of this style by reflecting and by freely varying these possibilities.[8]

Or, presumably, we can also remain unconscious of this style, in the sense that the world has this style whether or not it has been made thematic—that is to say, this style, this habitus of the whole, is "obvious." This is an important point, for it is not science itself, the pursuit of a method for constructing in thought the relations that make up the whole of the world, that first introduces the idea of an interrelated whole. The latter is already an accomplishment of pre-scientific life. Thus when Galilean science attacks the Aristotelian conception of the whole as a cosmos, transposing a world of "place" into the idealized "space" of Euclidean geometry, it is only Aristotelianism as a scientific construction of this whole in thought that is being rejected, not the originary, pre-scientific sense of the wholeness of what is. Not only is this "natural" sense of the whole not being rejected, but it remains something taken for granted, and represents a guiding motivation in what is an arguably Aristotelian understanding of the task of science as the construction, in a rigorous fashion, of how it is that this whole actually fits together. The difference between Aristotelian and Galilean science lies in a re-orientation of this sense of the task of philosophy as the science of the world in light of a new and powerful interpretation of the role of mathematics; but it is a re-orientation that takes place within a constant, fundamental orientation, always taken for granted, that is ultimately shared by both positions—that acceptance of the world in its fundamental, habitual style of presence.

And this re-orientation of Galilean science—thus its re-foundation (*Nachstiftung*) of the original Greek conception of science—occurs in the wake of a *success*, that of *applied geometry*:

> By means of pure mathematics and the practical art of measuring, one can produce, for everything in the world of bodies which is extended in this way [as a collocation of bodies in relative positions to one another], a completely new kind of inductive prediction; namely, one can "calculate" with compelling necessity, on the basis of given and measured events involving shapes, events which are unknown and were never accessible to direct measurement. Thus ideal geometry, estranged from the world, becomes "applied" geometry and thus becomes in a certain respect a general method for knowing the real.[9]

The geometrical, now successfully operating outside of its original boundaries, becomes a factor in the shifting sense of the confines of the world itself, of its developing style. But what does the world become, or how does it show itself in the wake of such a shift, one in which the geometrical plays such an important role?

The strange new idea of nature

To be sure, historically a great deal stood in the way of the development of this "general method for knowing the real." An entirely new set of physical concepts had to be developed, concepts that would not only be tangentially compatible with mathematical description, but could be mathematically constructed from basic principles. Anything less would be an unruly hybrid of approaches to describing physical phenomena, and not a genuine "science." In particular, the problem of the mathematization of the relation of time to physical events in dynamics was perhaps the chief obstacle to the full mathematization of the laws of motion, and could not be adequately solved until the development of mathematics in a direction in which geometry was no longer the dominant model. Also the very idea of causality itself and its description (thus the very idea of a "physical event") had to be freed from the once perfectly acceptable notion of causality as "the reason why something occurs," in order, ironically, for it to play a lesser role in physical description, thereby giving a certain precedence to effects over causes. The mathematization of nature, Galileo's great project, slowly transforms physics from a science of causes into a science of effects and the prediction of effects. Galileo's gradual transformation of the impetus physics of the Parisian school is a case study in the progression from (1) the traditional focus on the origin of movement in a "cause" to (2) the attempt to reformulate the very conception of what can count as an efficient cause by relying on immediate experience ("impetus" as a generalization of "muscular effort"), and finally to (3) a more purely mathematical treatment that surpasses both of these stages.[10]

However, the particular development of physical concepts is not what is decisive for Husserl. What is more important for him is the new conception of nature itself that can be seen to be emerging in this history, and above all how this conception finds a place in our experience of the surrounding world. Again, the point is not simply to understand what picture Galileo had of the world, and how this picture changed over the course of the subsequent history of science; the point is instead to ask how, through a Galilean *Stiftung*, a particular acceptance of a sense of nature takes hold, becoming the very confines of the world which we take for granted.

What is this "new idea" of nature? First it should be emphasized that for Husserl this idea finds its expression thanks to a *repetition* of the ancient ideal of an objective science of the world, an *epistêmê* meant to be the very embodiment in thought of all that lies within the horizon "world," or of the universality of being. The shift, or re-orientation represented by Galilean science revolves around what seems to have become, by the 17th century, an open possibility of renewed value for pursuing this objective science of the world by re-articulating, reducing, re-conceiving all phenomena in mathematical terms:

> If one is already firmly convinced, moreover, like Galileo—thanks to the Renaissance's return to ancient philosophy—of the possibility of philosophy as episteme achieving objective science of the world, and if it had just been revealed that pure mathematics, applied to nature, consummately fulfills the postulate of episteme in its sphere of shapes: did not this also have to suggest to Galileo the idea of a nature which is constructively determinable in the same manner in all its *other aspects*?[11]

To be sure, ancient philosophy did not simply outline a general claim regarding the possibility of objective science through mathematics, but also offered a specific model of a geometrically oriented (vs. oriented to "demonstration" in general, as in Aristotle) science of nature: Archimedean science. Galileo and the proponents of Parisian impetus physics such as Benedetti were all fully aware of this, and all considered themselves Archimedeans.[12] Nor was it simply applied geometry in general—its successful application in ballistics, for example—that showed that a mathematical physics in the modern sense was possible. Galileo himself was instrumental in first demonstrating the real possibility of such a science in his formulation of the laws of motion. On the other hand, Galileo himself did not operate with a full grasp of the implicit conception of nature that was being developed in the pursuit of this line of thinking—he does not even represent the high point of its elaboration in his own historical epoch. The explicit conception of nature as *res extensa*, and with that of the reducibility of all phenomena to the primary qualities of and events within a purely mathematico-physical space, was fully elaborated first by Descartes.

But the final elaboration of the idea of nature as *res extensa* already occurs in the wake of a shift within the self-understanding of what it means to pursue knowledge, or to capture the essence of nature in thought. The shift itself is not articulated, least of all by Galileo; nor is it something that has to do necessarily with the particulars of the Galilean world picture. It has to do much more with the *manner* in which this picture, or understanding of things, must be constructed in order for it to be a genuine theme for thought at all. What shifts is what above was called the claim of science with respect to what is meaningful and not meaningful. The way that nature is pursued in Galilean science, the manner in which it sets itself up in relation to nature in order to understand, is an expression of a fundamental shift in the manner in which this world is "for us," given as a theme in the experience of understanding. Thus, it is not simply that Galilean science employs mathematics in a more rigorous or universal fashion than did Aristotelian physics; nor is it that Galilean science is more empirical, or more experimentally focused. For Husserl, it is rather the case that Galilean science institutes, as definitive of the understanding of the task of thinking as such, the claim that the world is mathematizable, and *only thus* thinkable at all. The significance of this is not limited to an awareness of the limitless application of mathematics in physics, but amounts to a radical *rediscovery of mathematical being in the form of the world.*

Galilean science is guided by the idea of an ultimate cognition of the world, the grasping of its essence, and its claim is that such a comprehension must be mathematical. But this universal cognition can be actualized only by way of the "discovery" of the mathematical principles of phenomena—which in turn demands an ever more refined measuring of the world, an ever greater perfection of world objectification along an infinite scale of approximation that ultimately changes the very meaning of measurement itself. This implies that the essential relations that express what the world ultimately "is," relations which are, because ideal, constructed in systematic form, and which are visible only in the precise fixation of the objective shape of the common surrounding world in the language of mathematics, are not articulated in mathematical formulas simply because it is more efficient to do so. Rather, the claim is that this is the only way that the world can be meaningfully articulated as such; *any other articulation is a failure to grasp the essence of things*, it simply falls short of a comprehension of what is. Thus it is not as if one could, after having "translated" experience into the language of mathematics in order to understand, ever translate "back" into a language of experience untouched by the mathematical, as if one could return back to scientifically naive experience and still be able to find the appropriate concepts to be able to speak with meaning. Thinking is simply no longer at home in any such naive experience. And this is what begins, in Galilean science, to become the new, accepted confines of the world, or the new obviousness of things—what is obvious is that, if we are to understand, then we must engage the language of mathematics, because that is the very language in which one can say what there is to be said about the world.[13]

This is only one aspect of the shift. The other is more subtle, but more important, and it has to do with something that Husserl takes some pains to emphasize: the *strangeness* of this new conception of nature:

> If we adhere strictly to Galileo's motivation, considering the way in which it in fact laid the foundation for the new idea of physics, we must make clear to ourselves the *strangeness* of his basic conception in the situation of his time; and we must ask, accordingly, how he could hit upon this conception, namely, that everything which manifests itself as real through the specific sense-qualities must have its mathematical index in events belonging to the sphere of shapes—which is, of course, already thought of as idealized—and that there must arise from this the possibility of an indirect mathematization, in the fullest sense, i.e., it must be possible (though indirectly and through a particular inductive method) to construct ex datis, and thus to determine objectively, *all* events in the sphere of the plena. The whole of infinite nature, taken as a concrete universe of causality—for this was inherent in that strange conception—becomes [the object of] a *peculiarly applied mathematics*.[14]

This is the anticipation we have been looking for, the presentiment of the sentiments of the future. And it is a strangeness that is not confined to the initial historical emergence of the idea of nature that guides Galilean science; it is not a strangeness that disappears once the idea, or in this case the project, becomes familiar. Rather, the strangeness itself ultimately becomes a familiar strangeness, taking root in the acceptance of familiarity itself; and this rootedness is in turn progressively established through the continual advancement of the construction of nature as mathematizable being.

Growth and expansion is essential to modern science, and this is intimately related to the strangeness of the idea of the world with which science operates. The discovery of new principles requires the development of more complex methodological approaches to the investigation of phenomena, whether in the form of thought-experiments or actual physical experimentation of greater and greater technical sophistication; likewise, a parallel development of ever more systematic approaches to mathematics itself is required, leading to a level of formal axiomatization (and with that formal manipulability) of the sphere of the mathematical that had never even been attempted in ancient times. Since the "shape" of events themselves is more indicative of the mathematical essence of the world than knowledge of the causes of events, modern science develops into a systematic technization of induction, that is, of postulating the similarity of physical events that gradually leads towards a growing, and continually verified, generalization of types expressed in mathematical formulae. But none of this ever results in a final, fixed, and complete grasp of the world in a purely constructed deductive system of forms from basic principles. Mathematizability never stands as a fully realized project; the claim of science never ceases to be a claim "on its way" to being verified, even if it is always the point of departure, that in accordance with which thinking moves.

What remains palpable, and which is heightened in the absence of complete systematicity, is the strangeness of the world, to the extent to which the world could in effect even be identified as this strangeness. Or better: it is in its strangeness, we could say, that the idea of a rational order of the world becomes uniquely manifest; for the idea of a rational order of things is never "finally" instantiated for us in any given scientific articulation of the world. The "world," it itself as it is "in the end," becomes the pole of an infinite approximation; its being is that of an ideality that can be approached only within an open-ended pursuit of the technical mastery of inductive science, which has now assumed the mantle of *epistêmê*. We can still recognize in Donne's expression from *An Anatomy of the World* something of ourselves: "'Tis all in pieces, all cohaerence gone; / All just supply, and all Relation." The anatomy of the world, the confines within which our thinking has learned to move, is that of a world of which we can catch a glimpse only when our thinking begins with pieces, then through induction strives to project itself into the space of ideality. This, ultimately, is the great (re)discovery of Galilean science: the concrete presence of the ideal in the very heart of the real, the idea that the essence of the real is a hidden mathematical ideality. It is nothing less, as Koyré argues, than a new Platonism.[15] The ideality hidden in the very heart of the real is progressively brought out, realized in that *epistêmê* which undertakes a measurement of the world, not in order to fix the objectivity of things, but rather to fix their essential structures, their unique *ideal objectivity* that can only be expressed in physico-mathematical formulae.

This discovery of the world in Galilean science requires ever-more complex methods of measurement and, perhaps more importantly, of theory construction—already here, in germinal form, is the idea of a systematic philosophy that will take numerous forms from Spinoza to Hegel and, in some important respects, Husserl himself. From this point on, to theoretically orient oneself towards the world is understood as a striving to grasp a given *infinity*—not only in that the world itself, the universe of all that is, is "in itself" of an infinite extent (an idea that would find its time, vindicating Giodarno Bruno), but in that the appearance of the "true" world is only possible as the end result of an infinite *activity* that yields more and more precise approximations. The idea of nature as mathematizable motivates the development of methods of translating the remnants of a more immediate, intuitive understanding into a more precise "mathematical" understanding, whereby the age old activity of idealization already operative in pure geometry and the arts of measurement is generalized into a way of talking about the essence of the world as a whole. And again, the demands of this generalized idealization motivates the further development of mathematical language itself (algebra, but also the calculus, which Husserl for some reason does not cite).

This needs to be stressed: Galilean science, the "new Philosophy" (Donne), does not simply, or even most importantly, introduce a new picture of that natural world which, "all in pieces," European subjectivity then uses to interpret its place in the midst of things. With Galileo is inaugurated an *infinite project of mathematization* as the new understanding of the task of humanity, of the Greek philosophical ideal—and it is *this* task, understood as grounded in the insight of the mathematizability of the world, thus of the natural world as a certain type of infinity, that has become familiar to us.

How did such a familiarity take hold? How did the very strangeness of this idea anticipate our world, sensing the possibility of its emergence as a possible way of understanding? It did not take hold all at once; it was rather thanks to a shift within the movement of the obvious itself, a gradual displacement of a "common sense" world by the strange new idea of nature, thanks to which the expectation of an unproblematic connection between the world and human sensibility is gradually deflated. The hold of this idea is a function of its very openness, the postponement of any final word as to what the world actually looks like. Husserl emphasizes the open-ended character of Galilean science as a project, something that finds its counterpoint in the growing importance of verification.

Galilean science does not, as has already been stressed, introduce for the first time a mathematical picture or vision of the world; nor is it in itself a system of physics, a fully articulated theory of the world as a whole. Descartes, in his *Principles*, attempted to articulate and ground such a picture of the world, and in doing so cut his physics off from its own possibility as a realization of the Galilean idea, as Koyré has argued.[16] The very methodological stricture of clear and distinct ideas cuts off the very possibility of a theoretical approach to the world as a complex physical system that would at any given time leave the "whole" indeterminate, the pole of an infinite approximation. One can argue that Galilean physical science is the "beginning" of modern science in part because of its indeterminate character as a project—with all of the open ends that entails, since from the Galilean point of view there is no reason not to approach the "mathematical essence" of the world by isolating individual types of physical phenomena, leaving the interconnectedness with other phenomena indeterminate, in order to arrive at individual formulae expressing laws governing particular aspects of physical motions. This is an important consideration, because for Husserl the manner in which tasks are posed in this attitude is of decisive significance. The Galilean mode of posing questions, and of handling the results that emerge, exercises a powerful legacy for the further historical development of the tasks of science; above all, the Galilean beginning in its very indeterminacy nurtures the historical space within which the idea of science as an infinite task is first made possible, that is, where the points of departure for infinite tasks are always *local problems*.

Thus the significance of 17th century science for Husserl is not only that it has finally understood that the world is infinite, but more importantly that science as such begins to reveal itself as a kind of infinity as well. *The idea of reason is revealed to be the idea of an infinite task*, and in a manner that projects a particular style of the formulation of problems and the pursuit of method. Science no longer attempts simply to discern the justifiability of theory, to rationally demonstrate a given explanation of the world; theory is instead pursued from the beginning in accordance with an understanding of the theoretical as something set within the horizon of its own verification, which also means its ongoing perfectibility. Insofar as verification is something that is inherently open—not simply because it can be repeated, but also because in the very repeatability of verification the "truth" of theory can change as it approaches the pole of its own perfection—means that verification is also always already hypothesis:

> In the unending progression of correct theories, individual theories characterized as "the natural science of a particular time," we have a progression of hypotheses that are in every respect hypotheses *and* verifications. In the progression there is growing perfection, and for all of natural science taken as a totality this means that it comes more and more to itself, to its "ultimate" true being, that it gives us a better and better "representation" [Vorstellung] of what "true nature" is. But true nature does not lie in the infinite in the same way that a pure straight line does; even as an infinitely distant "pole" it is an infinity of theories and is thinkable only as verification; thus it is related to an infinite historical process of approximation.[17]

Technique and induction

All of this still does not really answer the question of how this idea of nature, and of the science of this nature, took hold. One can still ask: how is it that this historical process of approximation became a concrete possibility, or how did an infinity come to inhabit history as a motivating factor?

There are two considerations that need to be taken into account, one of which is expressly emphasized by Husserl, the other implicit in his text. (1) First let us consider what is emphasized by Husserl: science understood in the manner described takes the form of a *technique*. It is as a technique that science expresses its idea, or better: the founding (*Stiftung*) of the idea is the founding of a technique. The new task of science assigned in this founding, its "Galilean" task, is no longer the ancient task of "seeing," of achieving an insight into the real; the task is now to formulate a vision of the real within the horizon of the continuous verification of this formulation itself, the conditions of which are rigorously prescribed by the given theory as such. Such a science no longer approaches the task of theory directly from the question of what is real, but rather approaches the real with a manipulable, open-ended theory, which it seeks to perfect not only from the point of view of being a successful articulation of the real, but even as a *question addressed to reality.* This suggests that questions are no longer directly addressed to reality, but are prepared in advance, which means engaged as constructions in light of a multiplicity of possibilities for posing questions. Science itself becomes an art of posing questions, which in turn have meaning only relative to the infinite task of science. Therefore, science, as the technique of manipulating theory to approximate what is, takes the form of an attempt to come closer to that which it is always just beginning to learn how to question. The consequence is that the infinite approximation of science to the real never takes place in terms of a simple comparison of theory and experience; on the contrary, science in its essence *orders experience* by projecting what experience will or can become—i.e. what "will happen," or the way that the (true) world will appear within a theoretically ordered experience—by making use of a very sophisticated set of techniques for projection, including techniques of questioning.[18]

To be sure, this by no means implies that technique itself takes the place of experience, that the given is reduced to whatever constructions can be generated by the art of projection. Technique has its roots; it orders experience, but it also finds a place within experience; it enjoys a harmony governed by set of protocols in accordance with which whatever is constructed includes at the same time a necessary deference to what is given, even if the two become progressively indistinguishable in practice. Yet technique finds a home not by settling with the limits of experience, but by extending the possibilities of something that is always already at work in experience: the fundamental fact that the experience of the world occurs within a kind of anticipatory mode, which amounts to a prior having (*Vorhaben*) of the world in its possibilities. Technique thus finds a home in experience because, as a "task" (*Vorhabe*), it is already at home in experience:

> What do we actually accomplish through this technique? Nothing but *prediction* extended to infinity. All life rests upon prediction or, as we can say, upon induction. In the most primitive way, even the ontic certainty [*Seinsgewissheit*] of any straightforward experience is inductive. Things "seen" are always more than what we "really and actually" see of them. Seeing, perceiving, is essentially having-something-itself [*Selbsthaben*] and at the same time having-something-in-advance [*Vor-haben*], meaning-something-in-advance [*Vor-meinen*]. All praxis, with its

> projects [*Vorhaben*], involves inductions; it is just that ordinary inductive knowledge (predictions), even if expressly formulated and "verified," is "artless" compared to the artful "methodical" inductions which can be carried to infinity through the method of Galilean physics with its great productivity.[19]

Galilean science transforms the possibilities of an induction, or a *Vorhaben*, and a projection of meaning, or a *Vor-meinen* that are already constitutive moments of "all life." Science thus does not introduce an infinite horizon of inductivity to a world that had been closed to projections, thus undermining its closedness; it is instead a transformation of the sense of this openness of the world into something new, something this openness had not been prior to the taking hold of the idea of science, or prior to that historical event when science took hold, for itself and its own interests, of the open horizon of the world. This is important to keep in mind when one reads on to the next paragraph in Husserl's text:

> In geometrical and natural-scientific mathematization, in the open infinity of possible experiences, we measure the life-world—the world constantly given to us as actual in our concrete world-life—for a well-fitting *garb of ideas* [*Ideenkleid*], that of the so-called objectively scientific truths [...] Mathematics and mathematical science, as a garb of ideas, or the garb of symbols of the symbolic mathematical theories, encompasses everything which, for scientists and the educated generally, represents the life-world, dresses it up as "objectively actual and true" nature. It is through the garb of ideas that we take for true being what is actually a method—a method which is designed for the purpose of progressively improving, in infinitum, through "scientific" predictions, those rough predictions which are the only ones originally possible within the sphere of what is actually experienced and experiencable in the life-world.[20]

Thus when science, to refer to a much cited passage from the *Crisis*, "dresses up" the world in a "garb of ideas," it is in effect dressing the world up in the anticipations of life—though, to be sure, where these anticipations have been given a specific form, precisely in accordance with the sense in which the idea of Galilean science has given scientific life a new task. The perfection—as an infinite process—represented by science is not the perfection of an already available knowledge, if by that we mean the "ordinary" common sense knowledge of everyday life; it is rather the perfection of the capacity for ordinary life to *project for itself a world.* Thus the represented "world" of scientific understanding, even scientific understanding as a world of representations, is grounded, ultimately, in the productive movement of life itself; whatever distortions may arise, whatever misunderstandings and falsifications may result from the advent of modern science as an understanding of the world, all such obscurities must be understood as the result of possible risks inherent in the self-projection of life as such, and not simply a mistake. Obfuscation is, so to speak, natural for a life that projects meaning.[21]

I will leave the important topic of the lifeworld for Chapter Five, but already it can be seen that simply penetrating the "garb of ideas" to reach the world of life as "ground" will not be enough; the task at hand is to understand how the move from this ground to Galilean science as an understanding of the world occurs as the seizing, and also forming, in accordance with the idea of science, of a possibility already present on the ground, so to speak, of concrete life. And, more importantly, the task is to understand this movement precisely in such a way that we can grasp how it is that this movement, which is a movement of life, poses for itself the risk of self-obfuscation.

(2) The second consideration in understanding how the new idea of science and nature "takes hold" is not expressly emphasized by Husserl, but is nevertheless a key thesis that runs through Part II of the *Crisis*, and which has already been described as the "zig-zag" characteristic of these historical reflections. Implicit in Husserl's argument in these passages is that the availability of the historically given for reflection is determined in part by its formation as a *tradition.* It is only within the reflective zig-zag oscillating between our situation and that of Galileo within the movement of familiarity that the infinity of science as a historical phenomenon becomes truly evident; but it is only once the idea of this science is present as a tradition, and not merely as a response to a tradition, that its historical essence can be present to us in such a manner that we can say, reflecting historically, that the possibilities revealed here have a hold *on us.* Thus Galilean science finds its fulfillment as a tradition which, reflecting historically, we can bring to clarity with respect to its guiding idea; which means that only "for us" is the Galilean idea an effective meaning, a coherent unity of sense, and not simply a possibility in the making. And it is also only "for us" that the description of modern science as a "garb of ideas," or as the infinitizing through technization of the natural movement of a *Vor-*

habe, has any sense; just as it is only "for us" that method has become paradoxically indistinguishable from our connection with even the question of the "real."

And, finally, it is only "for us" that the idea of Galilean science as an infinite task is genuinely available at all, precisely because this idea is possible, as a real possibility, only when it takes the form of a tradition. This thesis will be considered in more detail in the next chapter on the *Origin of Geometry*, which deals with it explicitly. Right now, I only wish to emphasize that it is from "our" position, as the fruition of a tradition that begins but is not fully present in the 17th century, that *both* the idea that defines this tradition *and* the obscurity or distortion of this idea "take hold."

Galileo's "fateful omission"

The ultimate purpose of Husserl's historical reflections in the *Crisis*, as was argued above, is to secure a perspective on the present possibility of philosophy, and of science—for our times, our situation. The nature of this perspective is that it sets out from what can be described as the problem of the having-been-anticipated of the historical present, or from that anticipation that formed the possibility not of what this present could be, but of what it is. "Our" philosophy, "our" science, is the realization of the possibility of the idea of an infinite task taking root in life, a hold that was anticipated by 17th century Galilean science in its "grounding" or assignment of the task of understanding the world. And in this anticipation, in the formation of this project or *Vorhabe* of science itself, there is not only the origin of the idea of reason that has come to define the familiar outlines of our sense of the world; there is also what Husserl calls a "fateful omission" that has led to the incomprehensibility of that very idea, and by extension the incomprehensibility of the historical present. This omission is what interests Husserl the most, for it is the genesis, visible only in historical reflections, of our own incomprehensibility, of "our" crisis; and the fact that it is graspable through an historical reflection means that we can make our motivation to address this omission clear, for it is only given this historical sensitivity that we can have this omission as an *issue* at all.

What is this "fateful omission"? It is, it should be stressed, a completely natural omission, one that occurred not because Galileo made a mistake, or failed to be rigorous enough to warrant either the title of scientist or philosopher. It occurred because of what Galileo took for granted—the tradition of geometry, the art of measurement, and the style of the world. To be sure, these "obviousnesses" (*Selbstverständlichkeiten*) themselves had in turn made possible precisely the grounding of the idea of reason that Galileo set into motion. Their obviousness was the ground for the discovery of science as an infinite task, thus the modern renaissance of the ancient idea of *epistêmê*, of the knowledge of the whole in a unique form. But, precisely as obvious, they omit the necessity, or the task, of an inquiry into justifying the obvious—into making clear what is already clear, or into making comprehensible what is already comprehensible.

> It was a fateful omission that Galileo did not inquire back into the original meaning-giving achievement which, as idealization practiced on the original ground of all theoretical and practical life—the immediately intuited world (and here especially the empirically intuited world of bodies)—resulted in the geometrical ideal constructions. [...] For in the case of inherited geometrical method, these functions were no longer being vitally practiced; much less were they reflectively brought to theoretical consciousness as methods which realize the meaning of exactness from the inside. Thus it could appear that geometry, with its own immediately evident apriori "intuition" and the thinking which operates with it, produces a self-sufficient absolute truth which, as such—"obviously"—could be applied without further ado. That this obviousness was an illusion—as we have pointed out above in general terms, thinking for ourselves in the course of our exposition of Galileo's thoughts—that even the meaning of geometry has complicated sources: this remained hidden from Galileo and the ensuing period.[22]

The genius of Galileo, to quote a famous passage in the *Crisis*, at once reveals and conceals; it reveals the possibility of an open-ended, progressive exactness, an infinite perfection in understanding and projecting the world for a theoretical life, while at the same time it conceals the meaning-sources of the idea of this perfection itself. We are in the grip of this idea, but it remains strange to us; it is unclear why or how it has a hold at all. Galilean science is the revelation of the idea of an infinite scientific activity that ultimately leaves the sense of this infinity in obscurity; the advent of this science inaugurates a radically new understanding of the world that changes the very meaning of "world" for scientific thought, yet it is also, thanks to an omission of taking steps to understand, a continuation of a naiveté with respect to the ultimate accomplishments of world understanding. The omission firmly in place, as firmly in place as any "discovery," a peculiar naiveté is firmly cemented, one which has sustained the tradition of science for centuries—and remains as our connection to its idea, an unquestioned bond:

> Thus no one was ever made conscious of the radical problem of *how* this sort of naivete actually became possible and is still possible as a living historical fact; how a method which is actually directed toward a goal, the systematic solution of an endless scientific task, and which continually achieves undoubted results, could ever grow up and be able to function usefully through the centuries when no one possessed a real understanding of the actual meaning and the internal necessity of such accomplishments.[23]

This naiveté masks the crisis; but it is also a manifestation of the crisis. For naiveté cushions those who experience the hold of the idea of reason from those who experience an explicit loss of faith in reason characteristic of the age, but without being a defense. Because naiveté, in the end, says nothing; and thus reason, functioning in part thanks to a peculiar naiveté, "speaks" as little to the believer as it does to the skeptic; one operates as if the silence of naïve reason were not audible, the other is deafened by it. Yet silent does not mean absent; and it is the acceptance of naiveté which in turn carries with it the possibility of articulating what had been taken for granted, or what had been accomplished, "necessitated" within the historical movement of this familiarity itself: the naïve acceptance of the idea, its being a part of the world, primes both the believer and the skeptic for an awareness of the crisis of the idea, for the necessity of coming to terms with the idea as an historical task.

It is here that the peculiar one-sidedness of Husserl's historical reflections makes itself felt. Would not his entire presentation be more powerful if it were not limited to the idea of reason and science alone, but were instead sensitive to the sources of the Galilean gesture in other dimensions of the Renaissance conception of knowledge? For example, we could draw on the discussion of Galileo found in Ernst Cassirer, for whom Galileo was not simply a scientist but more importantly a humanist; for Cassirer, the "idea" that here finds its establishment and assignment as a task is not only the idea of science, but of humanity as such.[24] Galilean science is one expression of a philosophical conviction of the limitless potential for the perfectibility of human knowledge, guaranteed by an ultimate commensurability of the order of things and the order of thoughts. Furthermore, as a recent study has suggested, the thought of Descartes cannot be fully understood outside of the Renaissance heritage of the idea of the *perfection d'esprit*; to read Descartes while "forgetting" the echoes of the work of figures such as Pierre Charron and Charles de Bovelles would threaten to overlook distinctive features of Cartesian thought, in particular the notion of the self as an infinite project of perfectibility.[25] And is there not a kind of "forgetting" of the Renaissance in Husserl's *Crisis*, precisely where he turns to the Renaissance in order to gain a perspective on the problem of philosophy as a task?

Yes and no. One the one hand, Husserl's interests in Part II of the *Crisis* do indeed have a narrow focus, given that he interprets the genealogy of phenomenology as being almost exclusively tied to the emergence of the problem of epistemology or, more precisely, the problem of subjectivity in its modern, "epistemological" form. The logic of this emergence governs the entire exposition of Part II, including the sections that follow the discussion of Galilean science (§§16-27) which sketch the development of both the incomprehensibility of subjectivity and the transcendental thematic as its countermovement in the history of philosophy from Descartes to Kant. Here the possibilities for a rejoinder with the Renaissance conception of wisdom are few. On the other hand, the problem of subjectivity is understood by Husserl as a thread that will lead us not only to the possibility of phenomenology, but to a clarification of the current philosophical situation that, once accomplished, is in fact understood by Husserl in terms that hearken back to Descartes' original title for his *Discourse*: "le projet d'une science universelle qui puisse élever notre nature à son plus haut degré de perfection."[26] And it is here that we can see a deep kinship with what Cassirer would call the "Renaissance philosophy of man." It can also be seen throughout Husserl's notes and working drafts for the conclusion of the *Crisis*;[27] but more importantly, it is also clearly compatible with the conception of *Stiftung* discussed above: to be assigned a task, if it is fundamental, is to be subjected to a demand "to be" in a certain fashion. The perfection of the idea, the gesture in which the idea becomes "genuinely" manifest, is not external to human being, but the very formation of the latter.

Nevertheless, the focus of interest which, at least temporarily, leaves aside the broader question of wisdom in all of its dimensions has an important function. For it allows Husserl to clarify something that is very difficult to keep in focus: the question of the nature of naiveté itself, in particular the possibility that naiveté is not an obstacle to knowledge, but a necessary condition for knowledge. This rather paradoxical assertion will find its justification in the chapters that follow. For now, I would like to suggest that while Husserl's focus on the narrow problem of subjectivity may at first strike us as dry and abstract when compared to what could be a far more rich and engaging account of the Galilean *Stiftung* in the context of Renaissance science generally, it in fact puts us in a position to reflect on the almost invisible role of familiarity and naiveté, precisely by simplifying the medium in which it unfolds.

What Husserl has to say about the role and nature of naiveté, or more generally how an accomplishment of meaning can be oblivious to its own sense and still remain an accomplishment, even a developing one, cannot be exhausted by considering the text of Part II of the *Crisis*. To take up this question, we need to turn to another text of this period, *The Origin of Geometry*, which is the task of the next chapter.

Notes

[1] Crisis 23; Hua VI 20:26-28.

[2] Thus to some extent we can concur with the thesis of A.Gurwitsch: "It is the historical significance of Husserl's Galileo analysis to challenge and even to abandon the acceptance of science as an ultimate fact and to see in it a problem." *Phenomenology and the Theory of Science* (1974) 37. That is, we can agree as long as we interpret "challenge" and "abandonment" of ultimate facts not as a rejection of their validity, but rather the form that this validity takes once it has become something obvious.

[3] Crisis 24; Hua VI 21:10-20.

[4] Crisis 58; Hua VI 59:7-16.

[5] Crisis 26-31; Hua VI 23-29.

[6] Crisis 27-28; Hua VI 25:5-11. What is not emphasized by Husserl, but which one could argue is nevertheless decisive, is *instrumentation*. As M. Clavelin argues (*La philosophie naturelle de Galilée* (1968) 405-407), the development of instrumentation plays a key role in the notion of a "stablized experience" in modern physics: "Seule une expérience stabilisée, ordonnée, où l'essentiel aura eté soigneusement dissocié du contigent, pourra donc fournir à la philosophie naturelle l'appui dont elle a besoin." (405). Though, to be sure, as we are about to argue, what in one respect is stable can in another respect be the ultimate loss of stability.

[7] We should of course cite the Pythagorean tradition in this respect, for which "to be" was to be countable, and in that sense a "number." But also, and perhaps more importantly in this context, we should think of Fr. 1 in Heraclitus: even that which is beyond our actual ability to see, to comprehend, is not thereby beyond the measure of things; the horizon of what is, or of the "world," "all things," is itself the *horizon of the measure* (*logos*).

[8] Crisis 31; Hua VI 28:18-29:3.

[9] Crisis 33; Hua VI 31:2-11.

[10] Here we are following the presentation of the conceptual history of early modern science found in Alexander Koyré's *Galileo Studies* (1940), which is in many respects remarkably compatible with Husserl's discussion of Galilean science in Crisis/Hua VI §9. (Whether this could have been influenced by a visit Koyré paid to Husserl sometime between 1936 and 1937, which is mentioned by Carr in his Introduction (Crisis xix), but nowhere mentioned in

Schuhmann's *Husserl-Chronik*, is not something I have been able to confirm.) Another, more recent study that complements Koyré's interpretation is W. Shea's *Galileo's Intellectual Revolution* (1972); a contrasting work that de-emphasizes the Archimedean influence (that is, the idea of an essentially non-experimental mathematical physics) can be found in L. Geymonat's *Galileo Galilei* (1965).

[11] Crisis 33; Hua VI 31:17-25.

[12] See Koyré, *Galileo Studies* (1940) 21-38. Nevertheless, the "Platonist" Archimedes did not necessarily simply replace Aristotle as an authority. Cf. S. Drake, *Galileo at Work* (1978) 178.

[13] J. Klein: "The fact that elementary presentations of physical science which are to a certain degree nonmathematical and appear quite free of presuppositions in their derivations of fundamental concepts (having recourse, throughout, to immediate "intuition") are still in vogue should not deceive us about the fact that it is impossible, and has always been impossible, to grasp the meaning of what we nowadays call physics independently of its mathematical form." *Greek Mathematical Thought and the Origin of Algebra* (1934/trans. 1968) 3-4.

[14] Crisis 37; Hua VI 35:31-36:8.

[15] See Koyré, "Galileo and Plato" (1943); also cf. Cassirer's discussion in "Galileo's Platonism" (1944), where he makes the point that Galileo's Platonism, if we are to continue to call it that, is unique in the history of Platonism in that it would have to be a "physical Platonism," i.e. where the radical discontinuity between the sensible and the mathematical is rejected. The use of geometrical analysis in physics, itself proposed by Plato but limited to "hypothetical" demonstrations that never confuse the fundamental ontological distinction between sensible being and mathematical being, is given in Galileo an entirely new interpretation: "He [Galileo] accepted Plato's hypothetical method but he gave to this method a new ontological status; a status that it never had possessed before. In Galileo's science we find no longer a clean-cut distinction between the physical and the mathematical world." (292)

[16] Koyré, *Galileo Studies* (1940) 79-94.

[17] Crisis 42; Hua VI 41:28-42:5. Thus Newton's famous remark "hypotheses non fingo" can be seen as both an affirmation of the ideal as well as its distortion—it is an affirmation of the precision of the method, but a blindness to the inductive character of the ideal, i.e., its native infinity.

[18] We note here that this is as true of modern mathematics as it is of modern physics and, of course philosophy. This can be seen in the history of the famous "Hilbert questions" in the 20th century. See J. Gray, *The Hilbert Challenge* (2000).

[19] Crisis 51; Hua VI 51:12-24.

[20] Crisis 51-52; Hua VI 51:25-52:14.

[21] G. Brand, commenting on these passages, sees in what we have called the perfection of induction the defining character of the modern scientific "world-picture": "Alle Weltbilder, die es bisher in der Geschichte gegeben hat, waren in sich sinnhaft. Zum ersten Mal haben wir nun eine Weltauffassung, die den Sinn selbst aus der Welt zu verbannen sucht." *Die Lebenswelt* (1971) 27. Perhaps one should be wary of exaggeration—it may be closer to the truth if we were to say instead that every age is marked by a tension with its own sense, or its own "Sinnhaftigkeit," in each case different, and where in each case the tension may or may not ultimately lead to a break. What would then make the modern era distinctive is not only the experience of such a break, but the continual enduring of a mode or style of life in its aftermath.

[22] Crisis 49; Hua VI 49:23-50:7.
[23] Crisis 52; Hua VI 52:15-24.
[24] See Cassirer, "Wahrheitsbegriff und Wahrheitsproblem be Galilei" (1937); also L. Cahoone, "The Interpretation of Galilean Science" (1986).
[25] See E. Faye, *Philosophie et Perfection de l'Homme* (1998); also N. De Warren, "Philosophy and Human Perfection" (2001).
[26] Pointed out by Gilson in his commentary to the *Discours*. See Descartes, *Discours de la méthode*, texte et commentaire par E. Gilson (1925) 93, cited in DeWarren, "Philosophy and Human Perfection" (2001).
[27] Cf. Hua VI 275:5-276:6, and Beilage XXVI.

Chapter Four

The Origin of Geometry

The indifference of ideality and language

The *Origin of Geometry*[1] is an important text for the reading of the *Crisis* being presented here for two reasons. First, it is a text in which the historical character of the naiveté of Galilean science—indeed all science—is appreciated as both productive and obscuring; thus here one can again approach the question of what Husserl means when he describes the genius of Galileo as "revealing and concealing." Second, it is by way of an analysis of the *Origin* that the question of the motivation of phenomenological philosophy can perhaps be posed more clearly, and in such a way that brings the analyses of Galilean obviousness in §9 together with the theme of the life-world in Part IIIA, all the while keeping in view the fundamental question of the nature of inner history.

As the *Crisis* stands, §9 appears to be an early digression in Part II, temporarily interrupting an exposition that is for the most part oriented around the problem of dualism in modern philosophy. Dualism is an important theme in the historical development of psychology, and for Husserl it is a symptom of a growing incomprehensibility of subjectivity itself that is a key moment in the awakening of the motivation for phenomenological philosophy. Its explicit appearance in the history of ideas takes the form of the crisis of psychology, and in many ways the rest of Part II after §9 (with the exception of §15) prepares the reader more for Part IIIB, "The Way into Phenomenological Transcendental Philosophy from Psychology," than for Part IIIA, which is devoted to the introduction of the theme of the life-world. To be sure, these two "ways into phenomenology" have several important points of connection, chief among which is the paradox of subjectivity; but the theme of history, carefully introduced in such a nuanced fashion in Part I and in particular §§9 and 15, seems to be all but left behind. This is in part a function of the fact that the original plan for the *Cri-*

sis, following the lead of the Prague lectures, was to center around the failure of psychology to provide a coherent science of subjectivity; the deepening reflections on intentionality and history characteristic of Husserl's development of the *Crisis* in the years 1936-37, while not incompatible with this initial focus on psychology, nevertheless represent a different set of emphases. These two lines of reflection are never really brought together completely, another sign of the unfinished status of the manuscript.

One result of the incomplete presentation of the theme of historicity in Part II is that it is easy to conclude that, at most, what one has here is an argument that Galilean science "forgot" or "passed over" the lifeworld, an omission that was repeated throughout the history of science and which has led us to an impasse with respect to the question or problem of subjectivity. Why it is that we must re-construct this history for ourselves, why such meditations are necessary, is a question that seems to remain unanswered, despite indications that it is in fact *the* question at hand. At the very least, the reader is left wondering why it is that the problem of subjectivity cannot simply be tackled head-on; one might even muse somewhat ruefully, as does Husserl himself at one point in a manuscript from 1934, "what does the philosopher need with history?"[2]

Such are the vagaries of an incomplete text. This makes the *Origin* all that more intriguing, for in this manuscript the argument introduces the problematic of the lifeworld from within the question of history itself, thereby evading the impression that the thematization of the lifeworld by phenomenology is simply the response to a "traditional" or "historical" question of philosophy—the nature of subjectivity—as if this question had simply been handed down by preceding generations. The problem of history leads the analysis of the lifeworld in another direction, which I hope to make manifest by emphasizing several key themes from the *Origin*.

The opening paragraphs of the *Origin* articulate a conviction of Husserl's already stressed above: the only mode in which we can clarify for ourselves the philosophical situation of our times, and with that the possibility of a philosophical life uniquely our own, is a reflection on the historical essence of thinking and what is accomplished in the wake of its unfolding, transgenerational activity. Again, Husserl's reflections here are not meant as a species of historiography, or a direct, factual assessment of how it happened that we have the tradition that we have; rather, the aim of these meditations is to find a way to formulate the question of the possibility of philosophy, of where it is that philosophy should "begin." This is presented with characteristic Husserlian caution: in asking where to begin, we also ask why it is that we are motivated to pursue a beginning at all, especially given that the motivation to do so is initially so obscure and uncertain. Phenomenological philosophy, after all, is not the *first* beginning of philosophy.

Such a beginning requires a certain depth of self-understanding even to begin. Modern philosophy does not stumble upon itself, but rather begins in a kind of self-anticipation that takes the form of the question of its own possibility, or of its meaningfulness as such. Before such a philosophy begins, it must first attempt to find its own beginning; but that requires a certain depth of self, or a self that is in command of an interiority within which it can pursue itself, not so much as a given or even a project but precisely as a question. Husserl's argument is in part that such a depth of self, or interiority of reflection, can be genuinely opened only through an historical reflection—understood as a reflection that moves over that terrain which belongs to our sense of the historical, exploring what this sense has in store for us, or searching for the possibility of an unexpected wonder. Another way to express this is that philosophy is primordially an exploration of "origins," uniquely present within and thanks to the historical sense of the present. Thus: the historical reflection on the understanding of the world, thus of the meaning that the world has for us, is ultimately meant to make thematic the depths of meaning, or its origins.

And it is such an exploration of what Husserl calls the "depth problems" of the meaning of the world inaugurated by Galilean science that is decisive for the possibility of articulating within this interiority of historical reflection the problem of philosophy, of its beginning:

> Let it be noted in advance that, in the midst of our historical meditations on modern philosophy, there appears here for the first time with Galileo, through the disclosure of the depth-problems of the meaning-origin of geometry and, founded on this, of the meaning-origin of his new physics, a clarifying light for our whole undertaking: namely, [the idea of] seeking to carry out, in the form of historical mediations, self-reflections about our own present philosophical situation in the hope that in this way we can finally take possession of the meaning, method, and beginning of philosophy, the *one* philosophy to which our life seeks to be and ought to be devoted. For, as will become evident here, at first in connection with one example, our investigations are historical in an unusual sense, namely, in virtue of a thematic direction which opens up depth problems quite unknown to ordinary history, problems which, [how-

> ever,] in their own way, are undoubtedly historical problems.[3]

But why is *geometry* so important? We have already seen that geometry is one of the three things taken for granted by Galileo, and that its acceptance played a key role in opening the possibility of the discovery of modern science as an infinite task. One issue that still needs to be addressed in order to clarify this role of geometry in the presuppositions of Galilean science is the meaning of the *ideality* of geometry—or better, the issue of the origin of this ideality, and of the originary foundation (*Ursprung*) of the idea of science established by its founding.

What is ideality? The theme of ideality was already referred to briefly in the last chapter, in relation to the activity of *idealization*.[4] Idealization is already an infinitization; the ideal represents the limit of an ever-progressing process of "perfection," of "ever more" precision and exactitude. Taking for granted geometry as an already-accomplished idealization, Galilean science is the re-discovery, in a different form to be sure, of an ideality already partly articulated in traditional geometry. But what is required in order for us to possess such an idea at all? What is being exercised, or what had to have been exercised in order for this idea to become part of our cognitive universe? To answer such a question requires more than a description of how it is that the particular, the factual, is projected towards the infinite goal that marks the limits of its perfection. Such a description already presupposes, and in that way passes over what is distinctive about ideality in itself—namely its *indifference*. The projection of idealization would not be possible at all if there were not already in place an attitude oriented towards that which, in its being, is indifferent to factuality. Having in view that which is indifferent to the real—thus the "ideal"—is presupposed by any ability to project the finite into the infinite; it is an aspect of the ground upon which any projection of this type ultimately rests.

Pure factuality is the unrepeatable, the "here and now" that, in its passing, stands in opposition to whatever could be said to not pass, thus to remain the "same." Because the ideal is indifferent to this opposition, thus to its own tension with pure factuality, it stands within the envelopment of its own sameness—it is in this way repeatable as the "same" in every repetition.[5] It is not the same manifest as remembered, or as a lasting image that somehow captured the likeness of something that had once happened "here and now," or "there and then"; it is precisely as the same repeated *both* then *and* now, indifferent to the difference between the two, even indifferent to the fact or accomplishment of the repetition itself. An ideality, strictly considered, is something that only occurs "once," this "one time" being sufficient for all the times that its singularity shines in repetition. And if, following Husserl, one were to argue that "life" is something that occurs because of the coming to consciousness of the difference between past and future, thus within the movement of time, then one must also accept that this indifference of the ideal is, in an important sense, in turn an indifference to the movement of life, to its very manifestation.[6]

The indifference of ideality can take a number of forms. Of interest here is the fact that it is not the case that for Husserl geometry is the most basic, primordial example of the appearance of such an indifference. More basic, and presupposed by geometrical thinking, is the ideality of *language*. Language (words, sentences, speeches) in its actuality, in its being exercised by those who speak with one another about the world, is present, or given as a store of ideal objectivities that are indifferent to their factual manifestation. Husserl uses the example of the German word "Löwe" which, like any ideal objectivity, occurs only "once," indifferent to its own repetitions in sensible speech or writing. But this ideality of language, as a form of being-once (*Einmaligkeit*), is fundamentally different from that of geometrical objectivities; for language is an ideality that is submerged, so to speak, in the activity of giving expression to things in speech. "Language" is the manifest though submerged ideality of expressivity itself, irrespective of what is being expressed. It is not, therefore, an ideality that is present in the mode of a theme or an object, and this is one reason why it is not an ideality that is on the same level as that of geometrical ideality.[7] For this reason Husserl must distinguish between (1) the ideality of the objectivities of geometrical thinking from (2) the ideality of the language used to express such objectivities:

> But the idealities of geometrical words, sentences, theories—considered purely as linguistic structures—are not the idealities that make up what is expressed and brought to validity as truth in geometry; the latter are ideal geometrical objects, states of affairs, etc. Wherever something is asserted, one can distinguish what is thematic, that about which it is said (its meaning), from the assertion, which itself, during the asserting, is never and can never be thematic. And what is thematic here is precisely ideal objects, and quite different ones from those coming under the concept of language.[8]

Husserl argues that there is an ideality characteristic of "expressions," such as the word Löwe, as well as what is being expressed, which can be referred to as the meaning, or what is "meant by" the expression. Now, what is meant by an expression can itself be, as Derrida in his commentary to this text rightly points out, either what Husserl in *Experience and Judgment* calls a "bound" or a "free" ideality.[9] One example of a bound ideality could be the meaning of the word "Löwe," which can be translated into different expressions in different languages while always remaining the "same" meaning. But the "meaning" of the word "Löwe," even if it is ideal *as a meaning*, in another respect has nothing to do with anything ideal; what is meant is something that is hardly indifferent to the world, but is precisely some living creature within it. Through the expression and the meanings that come into play in its unfolding, thinking aims at what is neither an expression nor a meaning. Thus the kind of sense that speaking makes when one speaks meaningfully of a "lion" carries along with it, at least when speaking of lions, the reality, as it were, of a flesh and blood individual lion, for this is what is "meant" when one says "lion"—the full accomplished sense of such speech is ultimately tied to the intuitive givenness of an individual.[10]

"Free" idealities, on the other hand, are precisely meanings that are never bound to the intuitive presence of individuals. Here the accomplished meaning of the expression neither lends itself to nor finds its completion in the articulation of the givenness of that which does not belong to ideality; it never breaks its spell of indifference to the individually given. For example, in the case of logico-mathematical idealities, such as a propositional function, what is "meant" in speaking or thinking about them is always only an ideal object, indifferent to anything worldly. Another, more complicated example is meaning itself taken as noematic sense, where meaning, qua pure meaning, has been made into an object in reflection.[11] To be sure, even on the level of free ideality, one must make distinctions between "exact" and "inexact" essences, but none of this would detract from the fact that the idealities freed in this manner would be characterized as "once and for all," fixed with respect to their objectivity in an independent fashion.[12]

Language, however, represents a mode of ideality of a different type from mathematical or even noematic ideality.[13] To be sure, language too can be "freed," in the sense that it can be made objective in a reflective regard that takes it up from the point of view of pure grammatical form. But what is thereby freed is not a sense or meaning, but something within which the expression of sense finds an order, and which in turn does not thereby exhaust the phenomenological essence of expressed meaning. More, the coherence of meaning as such for Husserl does not have its genuine origin in language; rather, meaning ultimately has to do with the manner in which things in the world are given to consciousness. That this givenness can be expressed, and even that this expression can set the stage for more complex forms of givenness, should not lead us to confuse the theme of expression with that of givenness; the givenness of the world is not the same thing as its expressedness. It would thus be incorrect to say that for Husserl the world is given "in language," if by that is meant that the givenness of the given could be considered to be first accomplished within the fold of its expression, or its articulation in linguistic statements or propositions.

And yet—and this is one of the things that makes the *Origin* so interesting—nor is language a mere tool, an open window, or a fisherman's net used to gather together meanings that have already been constituted in "original" experience. Language is instead an accomplishment that plays a key role in opening the possibility of objectivity as such, whether ideal or real. And language, too, is "bound" to individuation, to things in the world; but not in the same way as other types of bound idealities, such as the meaning of the word "Löwe." Such meaning is bound insofar as it has to do with the manner in which things are given, which for Husserl means given to an intuition in which things present themselves "as such and such." Meaning, in other words, is ultimately that thanks to which the subject has a world; not that it is the form of the simple presence of the world, but rather insofar as the presence of the world is sustained by a consciousness of the significance of the given as the mode of its "appearance." Thanks to language, the subject "has" a world as well, but in a different sense than pure meaning alone—namely, it has the world precisely as something about which something can be said, about which it can *speak*. That is something very different from being conscious of the given in its phenomenality alone; for in language, what is accomplished (a sense for something) is not only bound to the givenness of the given, but to the speaking of a subject as well.

Thus language is a "bound ideality" insofar as it represents the manner in which the world is given as something to be spoken about, and not only experienced as a unity of sense. In an important respect this is both separate from, yet intertwined with the givenness of the world in meaning. Language is not the manner in which the world is given, but rather the manner in which we have this givenness of the world as something for which we in turn have words. To have such words—or to be in the possession of a language—means in part to be able to live in the space of an indifference to the world in its immediate givenness, even the immediacy of its sense, and from this vantage to speak about what it is that is given, in light of its givenness. Language is the free space, so to speak, just outside of this light of the givenness of things, within which we can find words to express both the immediacy of the world and its sense.

And language is, Husserl stresses, precisely *our* space, for it is here that we are present together, existing for one another in the manner of a *community*. Thus the world that we experience together, the "objective" world, should not be described in terms of the simple immediacy of an intuition, even if that intuition has always already found its way into meaning. The point is not just that the world has meaning. The surrounding, objective (not objectivist) world is always already given in an intuitivity structured by sense that can *in turn* be spoken about, which is thus always available thanks to a primordial expressability in principle of everything of significance in words:

> Everything has its name, or is namable in the broadest sense, i.e., linguistically expressible. The objective world is from the start the world for all, the world which "everyone" has as world-horizon. Its objective being presupposes men, understood as men with a common language. Language, for its part, as function and exercised capacity, is related correlatively to the world, the universe of objects which is linguistically expressible in its being and being-such. Thus men as men, fellow men, world—the world of which men, of which we, always talk and can talk—and, on the other hand, language, are inseparably intertwined; and one is always certain of their inseparable relational unity, though usually only implicitly, in the manner of a horizon.[14]

This horizon of language, of the commonality of the world as something shared among those who speak, is also, of course, the horizon of sense. The world that we share, that we "have," is a meaningful world. Yet to live through a meaning, to perform those noetic acts of cognition that have as their correlates "meanings," is not all there is to finding a word for something, or giving a sense the linguistic expression appropriate to it. One needs to distinguish between a narrow sense of having the world as something understood, thus as something "intuited" or "in view," and a broader sense of having the world as something not simply understood, but understood in such a way that it is there *to give*—to give to another in speech, to give in a pointing, directing, and leading not of one's own intuition, but that of another. This suggests an accessibility of the world characteristic of an understanding that is more than comprehension, in that it is uniquely generative of the comprehension and vision of others—in short, an understanding formative of the surrounding world, to recall a theme from Chapter One.

In this way, Husserl gives the outlines of a description of the primal phenomenon of *communication*—but communication is not here understood as a kind of wrapping one's thoughts ("knowledge") in a garb of expression that in some mysterious manner (such as an ability to interpret signs) is transmitted to the consciousness of the other. Language is instead being described here as a fundamental acquisition that has already secured, for those who speak, a horizon of being-related to things in a particular mode or manner. To be able to speak is not simply to be in a position to understand, but to fashion the sense of things, to present them to oneself and others in a generative fashion. In this way language becomes indissolubly intertwined with the lived experience of the surrounding world itself, not as a filter or mediation between world and awareness, but as a fixed horizon of a particular kind of possibility, a particular mode in which the givenness of the world is lived in the form of an already constituted communicability.

This needs to be made more precise. Meaningful speech—language taken in its full "concreteness," including its being intertwined with meaning—can be understood as a *double acquisition.* (1) First, it represents the acquisition of meaning itself. Consider Husserl's example, the word "Löwe." What a Löwe is, what makes a Löwe a Löwe, is something lived through or cognized in the perception of the flesh and blood Löwe; perception itself is the cognitive accomplishment in which "this" object, this X towards which the intentional experience of perception is directed, is given "as" something—a "Löwe." That the Löwe, or any object, is given "as something" means that in experience it is not simply itself, but present in the mode of having a certain manner of being-given. There is a necessary tension between the thing and its being-given, a tension which allows for the formulation of the phenomenological concept of "givenness."[15] Experiential knowledge of things is an apprehension of things in their *manner of givenness*—the meaning of something, the principle source of the meaning of a word, is how it is given to understanding, which is directed not simply towards the thing, but to what makes the thing "what it is." The result of such an understanding, or the experience of the given, is therefore not simply the de facto encounter with the particular thing, but the acquisition of an understanding of what that something is, or the manner of its givenness. This "manner" of the thing, whether clearly elaborated in all that is present in it or not, remains a permanent acquisition of consciousness that, once acquired, is independent of (or indifferent to) the actual encounter of the object "itself," the "flesh and blood lion (Löwe)." This yields the first dimension of the ideality of the acquisition of meaningful speech: the meaningfulness of things, their sense for an understanding that apprehends them, is a moment of ideality that stands in a basic but productive tension with the factical "here and now" of the given things themselves.

This ideality of meaning, however, is indifferent in another respect. Namely, it is indifferent to the act of understanding itself. This is definitive of its presence in consciousness as an acquisition. What a lion is, what makes a lion a lion, is a meaning that I can return to "again and again," thus it is something that can itself be approached as a kind of *object*. It is an object that remains identical and identifiable throughout the multiplicity of those acts of understanding that return to it, that make it a theme, that perform the intentional "what is meant" again and again. Such acquired meaning is thus "transcendent" with respect to these acts; for it is not the same, identical act of understanding that is repeated when I think again the meaning of the word "lion." When I return to something, my approach is not identical with the initial approach, for then it would not be a return, but the first time. What is the same, and which as a sameness runs through a repetition spaced apart by the irreducible difference between one act of thinking and another, is the correlate of these acts—the meaning, as something "transcendent." But the transcendence of meaning is, as André de Muralt points out, a "reduced transcendence," or an *irreal objectivity*—it is neither a real component of the individual act, nor is it a real component of the object, but a non-real transcendence.[16]

The acquisition of meaning (whether it is dormant or explicit, objectified) is inextricably intertwined with a second acquisition (2), that of the expression of meaning itself. Expression is an irreality as well, an objectivity that can be returned to "again and again," and it is ideal in a similar fashion as meaning, i.e. in the sense of a twofold indifference. The expression is on the one hand transcendent with respect to the act of expressing, or the act of making conscious the expression in its use—a transcendence that is felt when struggling to remember a word or phrase "on the tip of one's tongue." It is also indifferent to its own sensuous manifestations in the written mark, the symbol, or the physical sounds of speech; in such physical manifestations language is individuated, instantiated in a here and now that does not belong to language itself in its ideality. The word "Löwe" remains identical throughout its instances, the plurality of signs "repeats" in different temporal locations that which itself remains the "same" in its first and only time.

But language—the ideality of words and expressions—is not indifferent to meaning. Or, more precisely, language is not indifferent to a particular formation of meaning, a formation in which meaning has "come into its own" *vis-à-vis* its ideality. That is, language is not indifferent to the *theme of meaning as such*, thus the givenness that is characteristic of the manner in which meaning itself is given, intuitively present. That we speak about a world, give it expression, implies that we are oriented towards the meaningfulness of the world, even if all the while indifferent to what is given "in" such complexes of meaning. To speak, or to be able to speak, means that one is not limited to perception alone, to the mute fascination with the simple particularity of things, but is instead guided by the unfolding of a wider experience of "seeing," one that does not limit visibility to the being of the individual "thing." Language, to use another expression of Muralt's, is a testament to the ideality of meanings,[17] but that also means to the reality meaning has for us, where thanks to the fact of being in the possession of speech we have always already taken a step towards making meaning a theme in and for itself. Or, to use an expression from the *Origin*, language is not indifferent to meaning because language is the very embodiment (*Sprachleib*) of meaning.

For this reason phenomenology must begin with a reflection on language. Not, however, in order to explain this strange embodiment, which involves the peculiar notion of an ideal body, but precisely because it is in language that something like a phenomenology, the making thematic of essences taken as an independent region of investigation, has already been achieved. To be sure, language itself does not *belong* to the region of sense, for the ideality of language is not the ideality of meaning; but the question of language is of unique importance for phenomenological philosophy, insofar as language represents a secure access to the meaningful as a theme for its own sake. Thus phenomenology makes language a theme only because what it seeks, the understanding of meaning as essence, is already "embodied" in the language of world-understanding—in particular, for Husserl, in the language of the sciences. It is that which is "in" language already that phenomenology seeks to make its own; though in doing so, it is not thereby leaving language behind, as if it were attempting to take meaning out of language and turn it over to a phenomenological seeing that would no longer require embodiment in words. On the contrary, the bond between phenomenological thought and language is irremovable. Phenomenology relies on language as the capability to find words for things, even the phenomenality of things; expressions are always already the embodiment of a nascent phenomenality. A successful phenomenology, or thinking that is exclusively oriented towards the ideality of sense, can do nothing more than confirm this basic proto-philosophical essence of language. Language is thus, ultimately, to be trusted; what the phenomenologist has to be attentive to is the ability to see, with insight, that which language has allowed to be expressed.

Of course, such an account of the place of language raises a number of questions, and ultimately points us to fact that Husserl never moved very far beyond the position of the *Logical Investigations* on the problem of language, in particular on the theme of its ideality. And the problem of the ideality of language is not something that can resolved here.[18] But it is important to be aware of it, because the ideality of language as an acquisition relativizes the achievement represented by geometry. Geometry is important for Husserl because it is the emergence of the theme of ideality in a manner decisive for the origins of philosophy. Greek philosophy for Husserl is premised on the growing awareness of the significance of the discovery of mathematical, ideal being, and the history of philosophy is in turn the continual re-discovery of this sense of ideality in ever more radical forms. But, as we have already hinted, the origin of geometry occurs within an horizon of human experience, not necessarily historical, that has already been irreversibly marked by the advent of language, which means that geometry is by no means the first emergence of ideality as an objectivity within which human understanding moves. If the origin of philosophy is the discovery of ideality, it is not the case that the origin of ideality is the discovery of philosophy.

Tradition and the problem of the "first acquisition"

At this point one might ask why Husserl does not begin with the origin of ideality as such. Why is the problem of the beginning of philosophy, and the task of making this beginning once again operative in and for the contemporary philosophical situation, located not in the initial acquisition of ideality for humanity, but so to speak only in its wake?[19] Though such a question may at first seem vague, it is in fact very important, for it bears on the relation between the theme of the lifeworld and the question of the beginning of philosophy that is pursued in historical reflections of this type. In fact, one could argue, a justification for situating the problem of the beginning of philosophy in the origin of geometry and not in the origin of language can be established by a careful reading of the text of the *Origin.*

The key point to make is that the possibility of philosophy as a real possibility is not present in the acquisition of language, nor even of ideality in general, but in the acquisition of geometry specifically as a body of *sci-*

entific knowledge. And this knowledge, though on one level present only if preserved in language, is genuinely given only in the form of a tradition. The implication is that the "language" of geometrical thinking is not language pure and simple, itself more original than any tradition, but precisely a traditional language suited for thinking about geometrical ideality.

Tradition is, of course, an historical concept; more importantly, it implies a sense of history that does not limit the presence of the past to testimonials preserved in documents. A reflection on the sense of history that is present in the form of a tradition is not a reflection on the sense of history that lies at the basis of historiography. Historiography seeks to reconstruct past events by basing itself on source documents; these documents testify to the "past" as a totality of events only if the historiographer takes pains not to allow the contemporary understanding of such events to influence the reconstruction. This is an ideal, to be sure, and one which is not unjustified, since there is an important sense in which the historical can become an object only given a turn away from the present, precisely in order to allow the past to come to light "on its own." We have already seen how Husserl's historical reflections are fundamentally different in their manner of thematizing the past, precisely because he understands such historical considerations as the means by which we who occupy the historical present reflect, not on the past as such, but above all on our own situation. The concept of tradition belongs precisely to this unique historical reflection that has a built in "zig-zag" between a past that sheds light on the meaning of the present, but which itself can only come to light given its interpretation in light of the present. It is, in effect, a reflective engagement with the tradition, with the past as it is refracted through what is of concern for us today. But what precisely is a "tradition," and why is it important for the analyses of the *Origin*?

Tradition can be described as the actual presence of history. It is the manner in which the historical past is an organic part of the surrounding world. As such, tradition is more than memory, because it is more than the ability to recall a moment in time; in fact, it is not a memory at all, but the remains of the past, aided by memory but not completely dependent. For this reason Husserl treats as equivalent the expressions "tradition" and "acquisition." A tradition is an acquisition, which means that it is an accomplishment of subjectivity that remains a permanent feature of the communal world. Thus tradition is above all a kind of product, one which exists for us in such a manner that we are aware of its historical character. Being aware of the historical character of our common cultural world is precisely an awareness of its having been produced, that it stands before us as an accomplishment of generations that came "before." Our very grasp of this historical (and not merely temporal) "before" originally takes the form of the sense of accomplishment that defines the basic style of the enculturated world.

Tradition is in this way a fundamental characteristic of culture, for it expresses at once its historical *and* productive character. But it is an historicity and productivity in and on the basis of which we live, and it is something with which we are thoroughly familiar—this is what it means to say that it is a part of the surrounding world. Tradition is an historicity and productivity that is thoroughly taken for granted; furthermore, it represents an obviousness that not only colors the sense we have of the past, but of the future as well. Insofar as it has this character of familiarity, a tradition is close to us, and ultimately something alive; to be "in" it is to be part of its forward movement, which means an engagement with the future. The past, as has already been argued above, fashions not only those possibilities that we, in the present, have ourselves made actual, but also in part those of our future. A tradition is as much the production of possibilities as it is the production of those actualities that are the concern of the historiographer. Standing within the familiar pattern of a tradition, it is "obvious" that certain things are possible, that certain things can be done, pursued, accomplished. Such obviousnesses (*Selbstverständlichkeiten*) can either take the explicit form of the obvious legitimacy of a certain conviction or idea, or can remain latent in the form of certain tendencies of which we are barely aware.[20]

Since we are thoroughly familiar with tradition, since it is in turn our very familiarity with our own productive historicity, we can be said to be knowledgeable about the tradition within which our communal life moves, for this tradition is an organic part of our surrounding world. Nevertheless, such knowledge does not take the form of a registry of facts; in the end we always know the facts of our history only incompletely. However much we may lay claim to our own history, it nevertheless remains characterized by inherent tendencies of ambiguity, distortion, and loss. But however much we may lose our history through misunderstanding obliviousness, we never lose history as such; however little we may actually understand the content of the cultural tradition to which we belong, and in which we live, we nevertheless cannot fail to be in possession of its "sense," its presence as that which forms the world as our common cultural universe. Thus, Husserl argues, even in the absence of an explicit knowledge of the tradition, there is an implicit knowing (*Wissen*), a primal understanding that always remains as a subterranean grasp, as it were, of the tradition in its living unity. Such an understanding is implicit in commonplace expressions. What is obvious about tradition is not what is explicit; what is obvious is obvious precisely because it is implicit, hidden from view. Active seeing and inquiry naturally pass over what is contained in obviousness in search of another object, another answer. As was the case with the analysis Galilean obviousness in Part II, Husserl's strategy in the *Origin* is to make what is implicit in the obviousness of the geometrical tradition the object of a new inquiry that seeks to transform what is implicit in our common understanding of tradition into a fully articulated as well as *evident* knowledge.

Yet such an historical investigation is possible only because history is present in such a manner that it lends itself to an inquiry of this type, that is, only because it is available as something to be questioned, something *there* to be brought out into the light of its own self-evidence.[21] And it is tradition—an obviousness possessed by everyone whose life has been somehow stamped by the "traditional"—that represents precisely such a mode of historical presence. Furthermore, and this is something we should not lose sight of, such an inquiry is possible only in the form of philosophy, if we understand (provisionally, to be sure) by "philosophy" the questioning of what is naturally unquestioned, or obvious. Only when obviousness itself has become a problem, a sphere accepted as questionable, can historical reflections of this type in turn become truly possible. Thus the project of the *Origin*, which is to pose the question of the "first geometer," the subject of the historical moment of the origin of the geometrical tradition, all the while remaining *within the horizon of the sense or meaning of traditionality as such*, can for this reason only be described as a "philosophy of history."

Husserl's inquiry thus begins with "the most obvious commonplaces"—chief among which is the fact that we possess geometry as a tradition, which means that we possess geometry as an acquisition. More precisely, geometry is an acquisition that was not simply handed down, but was developed as it passed through the generations that came before. There have been earlier forms or formulations of geometry that have since been perfected or replaced by other forms. The acquisition of geometry that we possess is a complex structure made up of whatever place or non-place was found in a developing whole for a multiplicity of individual creative accomplishments. And the "place" that any given accomplishment has within the "total acquisition," which is the tradition, is defined not only in terms of its meaning or importance, but by its mere presence as a predecessor. Simply by being that which has come before, the "preceding" acquisition always remains as a moment upon which the next moment builds, even if the new formation of meaning is an explicit rejection of what had been accepted as accomplished by the former. In this way, every "new" form of geometry, as Husserl says, repeats this reference to what had come before.[22] This reference enables Husserl to formulate the problem of the "first" acquisition: tradition is forward-moving; each creative act that finds its place the totality is an addition to a series of additions that in turn must, "obviously," refer back to a first acquisition. But just as obvious is the fact that there is a clear limitation of what could have been acquired "at first," a limitation that relativizes the originality of the first geometrical act:

> Every spiritual accomplishment proceeding from its first project to its execution is present for the first time in the self-evidence of actual success. But when we note that mathematics has the manner of being of a lively forward movement from acquisitions as premises to new acquisitions, in whose ontic meaning that of the premises is included (the process continuing in this manner), then it is clear that the *total* meaning of geometry (as a developed science, as in the case of every science) could not have been present as a project and then as mobile fulfillment at the beginning.[23]

The idea that Husserl first touches on in this passage is that the being of geometry *as a tradition*, thus as a "mobile fulfillment," is *necessary* in order for the full self-evidence of geometry to be possible. There is thus no "first" geometry, only "the one" geometry, which can be given or acquired only as an *already formed tradition*. It is as if, in order for the experience of the ideality present in geometry to occur at all, it is necessary that human consciousness already be in the possession of a depth within which, like the question of philosophy itself, mathematical existence can become something "given" as a genuine actuality. Husserl in fact makes such an argument, though with some important reservations; we should step back, before this manner of speaking, as he puts it, gets too "overblown." In particular, we should be cautious, lest we forget that what is present in and as tradition in the case of geometry is ultimately indifferent to the living, forward motion of tradition, and that this will be definitive of its self-evidence. The *content* of geometry is, after all, *ideal*. In its *original* self-evidence, in the sense of *genuine* self-evidence, geometry is nothing that "moves forward," however much the life within which such evidence is first acquired is such a movement.

The primacy of the I and the proto-community of geometers

As can be seen, Husserl's entire analysis in the *Origin* turns on how to understand the problem of the "first geometer." The first geometer is precisely the one for whom geometry is not a tradition. As the first, the first geometer is the one for whom ideality is not already articulated as an object for thinking, inquiry, and construction, but who nevertheless inaugurates just such an articulation. The "first" acquisition lays the basis for a style of thinking that has as its object mathematical existence, or ideal objectivity. As was stressed above, this does not represent the first *appearance* of ideality in human history; language is more primary, presupposed by geometry itself. As such, language is not only something already discovered, but an ideality that has become part of the world; and in so doing it is an ideality that has opened up a dimension within which humans are able to speak and express themselves to one another about the world. Geometry, too, precisely in becoming a tradition, is an ideality that has become a part of the world, again in the sense of a dimension within which we are able to think and to understand—and for Husserl, it is *this* advent of ideality as a space for thinking that is the origin not only of science, but also of philosophy.

Nevertheless, there is a shift of orientation in the presentation in the *Origin* with respect to this question that is telling. The question of the origin shifts from the original acquisition of ideality by the first geometer to the *traditionalization* of that acquisition. Husserl's problem is not how it is possible that ideality could have been discovered by a concrete individual who had had no preparation for such a discovery; nor is it a question of imagining the act of will that enabled the first geometer to peer behind the Heraclitean veil to catch a glimpse of an eternal perfection, an experience that would have a lasting influence on the generations that came after him. Instead, Husserl remains within the scope of questioning the sense of that which we already possess, namely the implicit knowledge of tradition itself; and from within this perspective the question automatically takes on a certain form. Namely, the question is what this obvious necessity of a first acquisition has to tell us about the manner in which ideality became, in the form of geometry, present as a tradition. Or: what does the obvious necessity that there was a first geometer tell us about the ultimate *nature* of this traditionalized acquisition? Or again: if we accept that the "cultural world" is precisely an "acquired world," then what does the necessity of the first geometer allow us to ask about the nature of the relation between *ideality* and *world*?

Not that the question of the geometer's original grasp of the ideal is simply a non-issue. Quite the contrary. It is only that this accomplishment can only be thematized from the perspective of an inquiry into its inner relation to the tradition. Husserl argues that it is legitimate to trace back, through a regressive inquiry into what is implicit in the sense of the tradition from within the historical present, the movement of the tradition "back" to an individual who, as the "first," inaugurated the tradition. The argument here is that the beginning is not indifferent to, but requires the individual; ideality must "first" occur within the individual "I." The "I" is the entrance, the only possible entrance, through which ideality can become meaningful for life. Thus we can say that the project of the *Origin* is a description of the structure of this entrance of ideality into history thanks to the I, one that to an extent neglects the problem of its entrance into the sphere of the I as such. And in fact the primacy of the "I" is not fully explained in this text, nor in the *Crisis* as a whole, though it is a key concept that will again be an issue later when we take up the transcendental reduction in Chapter Six. In the *Origin*, Husserl immediately moves on to an analysis of the objectification of the ideal that proceeds from within the "I" to the intersubjective community, an objectification which eventually takes the form of traditionalization.

The best way to understand the pre-stages of traditionalization is to think of a multi-layered process of concretization. This implies the making concrete of something that, in its most original self-manifestation, is nothing in itself concrete. For ideality, in its purity, is after all indifferent to the concrete; thus its concreteness is not a given, but must first be made, fashioned within a perspective that lends it the fastness of the empirical. At the most basic level, the concreteness of the ideal is its individuation, its being assigned a place within the movement of a thinking, thus being presented within a temporality that is not its own. This means that already, at the very beginning, there is a kind of impurity at play with respect to the self-evidence of the ideal; but it is an unavoidable impurity, for in order for the ideal to be meaningful, it must be given a form in which its indifference is something more than mere silence.

Furthermore, in another kind of shift that will be important for the phenomenological reduction, this concretization of the ideal takes place along with the continuous concretization of the "I" itself. The originary, pure "I" can be said to be left behind along with the originary indifference of the ideal; both undergo a formation within a concrete life in which they become intersubjectively available objectivities, until they are finally concretized within traditionalization. On one level, this concretization takes place within language. Language is that space in which we can speak of the world; in language, the world is available in the form of things for which we have words, and in which we are in turn available to one another. For among the things of which we speak are included our own ideas, emotions, and experiences. This means that whatever was originally present in its self-evidence "for" the first geometer already falls within this horizon of shared expressivity—whatever the geometer has, as one who experiences, thus as an "I," has always already undergone a process of being placed within a common intersubjective horizon of sense.

There is another, deeper level of concretization, however, both of the "I" and of the ideality of geometry. First it is important to be more specific about what is meant by the phrase "concrete." Something that is concrete is something that is fixed, in the most general sense; as fixed, it is available as either a subject for inquiry or as the content of a thought or an expression. "Concrete" thus means "individuated," present as a pole of articulation; but this does not mean static, unmoving, or unchanging. Concrete need only mean a play between active and passive; by no means is the concrete somehow impenetrable, or immobile. A concrete situation, for example, is hardly immobile, for it is a situation only because it is in motion; but it is "concrete" only if it is given definition by an ordered unfolding of events that are fixed by a passivity that gives them enough rigidity so as to set an activity into relief, thereby allowing for an orientation within a stable (but not necessarily permanent) context. All objects in the world bear the stamp of this phenomenal character of passivity—the book is "where I left it," the well-practiced hand can "find" the proper note on the piano, the lover "returns" to the sweet memory again and again. A concreteness in this sense of a play between active and passive is also present within the individual subject, precisely as an individual, pre-communal "I." And it is here that ideality finds its first concretion, its first "objectivity" as the objectification of its primal immediacy:

> Let us reflect. The original being-itself-there, in the immediacy [*Aktualität*] of its first production, i.e., in original "self-evidence," results in no persisting acquisition at all that could have objective existence. Vivid self-evidence passes—though in such a way that the activity immediately turns into the passivity of the flowingly fading consciousness of what-has-just-now-been. Finally this "retention" disappears, but the "disappeared" passing and being passed has not become nothing for the subject in question: it can be reawakened. To the passivity of what is at first obscurely awakened and what perhaps emerges with

> greater clarity there belongs to the possible activity of a recollection in which the past experiencing [*Erleben*] is lived through in a quasi-new and quasi-active way.[24]

Note that this analysis presupposes the auto-constitution of the "I" as *time*. As temporal, any accomplishment of the "I," any experience in which something is given in self-evidence, occurs within a relation to a passivity that can be brought to bear on what is being experienced—the original living through of an experience can be re-shaped, as it were, re-understood, by a quasi-living-through of a "past" experience that emerges out of the passivity of the temporal background of the I. Original understanding, the originary grasp of a given, is thus not an isolated act, but can be enriched by an equally original repetition that is available only because of this passive givenness of the past. This enrichment bears directly on the successful apprehension of the identity of what is originally present. This identity, in its individuation, is an example of a play between passive and active. It is in the mode of a passive "being-there" as a possible something to which I can return; but it is also in the mode of being-presented in an active articulation or "production" of what is "there" to be understood, an active grasp of what is now there that is in part founded on the enrichment of the given past, thanks to which the given is an identity in self-evidence. Let us continue the above passage from where we left off:

> Now if the originally self-evident production, as the pure fulfillment of its intention, is what is renewed (recollected), there necessarily occurs, accompanying the active recollection of what is past, an activity of concurrent actual production, and there arises thereby, in original "coincidence," the self-evidence of identity: what has now been realized in original fashion is the same as what was previously self-evident. Also coestablished is the capacity for repetition at will with the self-evidence of the identity (coincidence of identity) of the structure throughout the chain of repetitions.[25]

Again, this individuation of the original self-evidence of geometry occurs within the life of an individual I. It appears precisely as the "original" production, in thought, of the self-evidence of geometry. This concretion is not simply a fledgling preservation, however, but is a key moment in the development of the *theoretical interest* in geometrical ideality. An interest, and its development into a project, must be accompanied by an individuation of that towards which the interest is directed; this makes possible the development of the interest as a project of working towards a now *given* goal. Both individuation and interest dovetail in the idea of an originary experience that is the "origin" of both. However, though individuation is necessary, it is not sufficient; it is a formal condition, and does not supply the specific impetus for the motivation or interest in geometrical ideality as such. And this specific origin of the interest in ideality remains hidden when Husserl turns to the question of intersubjective communication. Based upon the prior functioning of empathy and language, whatever is accomplished by the original geometer "alone" can be communicated to others—not in the form of simple information, but as something which is, as Husserl stresses, "*actively* understood by the others":

> In this full understanding of what is produced by the other, as in the case of recollection, a present coaccomplishment on one's own part of the presentified activity necessarily takes place; but at the same time there is also the self-evident consciousness of the identity of the mental structure in the productions of both the receiver of the communication and the communicator; and this occurs reciprocally. The productions can reproduce their likenesses from person to person, and in the chain of the understanding of these repetitions what is self-evident turns up as the same in the consciousness of the other.[26]

Husserl's implicit claim is that the shift of the question from the original acquisition "in itself" to the concretion of this acquisition on the level of intersubjectivity passes over nothing essential. There is a structural parallel between what happens within the originary accomplishment of the geometer and what, in communication, occurs so to speak on the "outside"—outside of the origin, but still inside subjectivity. The other "understands," because the other also performs the originating act of grasping ideality; together, the companions pursue their developing interest in ideality. And this grasping, thus this originary self-evidence of ideality, is in turn enriched within a sphere of passively individuated presentations, though now such individuations are not limited to the past of the individual geometer, but also include individuations of ideality accomplished by the others. Thus the identity of geometrical ideality becomes articulated within a coincidence between the active production of the given individual geometer and the passively constituted productions of the "others" that are available to him, thanks to his place in a community. For in empathy and language I "have" the experiences of other subjects, though not in the mode of "my own," but as presentifications to which I stand in a particular type of passive-active relation.

This identity of ideality, however, obviously presupposes the presence of an interest in ideality. Identity here requires a focus that gathers together the genuine achievements of each individual into the intuition of ideality in its original self-presence. Or better, what is presupposed is the presence of such an interest within a *proto-community* characterized by a reliance upon the subjective accomplishment of clarifying self-evidence, precisely as something pursued, and only pursued, by each individual. As such, the "objectivity" or intersubjective availability of these productions is clearly limited in its scope, for it is dependent upon the originary rediscovery, *within each individual*, of the possibility of geometrical self-evidence, which is then confirmed (or concretized) in communication with the others. Husserl is not clear on precisely *what* is communicated between the first geometer and the proto-community of geometers (and I would argue that such clarity is not possible), but the logic of his presentation would suggest that it could only have been a kind of showing, or a *prompting* on the part of the first geometer for his companions to turn within, to orient themselves to a possibility of which they had previously had no idea. "Seeing" is here originally intertwined with the coaxing of an interlocutor, or of the promises of the unexpected which always accompany discourse.

Writing and the "seduction of language"

Whatever objectivity geometry wins as a result of the exchanges within the proto-community of geometers, which are oriented primarily around showing and seeing and only secondarily around documentation, it is an objectivity that remains dependent upon the continuing individual existence of the geometers. At this level, it is still necessary to contact a certain particular group of individuals, the "original" geometers, in order to "see" geometry being done, which means that geometry is here at most a quasi-objectivity, one achieved only from within the awakened, active interest in a subject (or group of subjects) who makes the original evidence of ideality a theme and concern:[27]

> Now we must note that the objectivity of the ideal structure has not yet been fully constituted through such actual transferring of what has been originally produced in one to others who originally reproduce it. What is lacking is the *persisting existence* of the "ideal objects" even during periods in which the inventor and his fellows are no longer

> wakefully so related or even are no longer alive. What is lacking is their continuing-to-be even when no one has [consciously] realized them in self-evidence.[28]

The "persisting existence" of ideal objectivity is constituted within another, very different relation between passive and active than had been the case either with the original geometer or even among the proto-community of geometers. This is, of course, the relation between passive and active in *written* communication. Understanding a written sign is a reverse image of the priority of activity over passivity that had marked the first two stages. While up until now it has been a question of an activity enriched by passive association, in the case of writing it is a question of activity itself being passively "called forth," or "awakened." This includes the possibility of what Husserl in these pages will call "reactivation," where the subject, prompted by a linguistic sign (whether written or verbal), performs the originary intention that would bring to original self-evidence the meaning signified by the sign—that is, where one would thus reproduce the meaning for oneself, in the active sense of bringing it to an originary consciousness. Such an act, one could say, would, in the example of geometry, admit one to the circle of the proto-community of geometers, thus to the community of those who are capable of performing the originary production of ideality in self-evidence. "Writing" would then make geometrical ideality "more objective," precisely to the extent to which it would open the inter-personal community of the proto-geometers to a world of strangers, to those who do not know, or who have never been instructed by, the original geometers.

Yet reactivation is not the primary form of the objectivity of meaning in written or even spoken language. The linguistic sign is not only something that offers the opportunity to partake in the original accomplishment of an interest; it is also, and primarily, the *embodiment* of the object of interest, in this case geometrical ideality itself. Written texts become the place where one returns again and again to access geometry, to such an extent that it begins to become counter-intuitive, or at least moot, to insist that the meaning of geometry necessarily lies outside of these texts; they hold the sense of the geometrical all their own. The result is that, in writing, ideality is present in a form that does not require an active understanding, or genuine production of meaning, even in order for it to be understood; the embodiment of ideal objectivity as a genuinely "persisting existence" means that its presence is no longer fully dependent upon the active understanding of the subject. To be sure, this does not mean that it is not dependent upon understanding in general, or an activity of some kind on the part of an understanding subject thanks to which access to the content of geometrical writings is maintained. The argument is rather that, with writing, the way is now open for the possibility of developing a unique kind of *passive understanding*, a mode of geometrical knowledge that had had no place, much less a function, within the proto-community of the original geometers.[29]

If one speaks here of an "embodiment of meaning," it is only in order to avoid the impression of a mere passivity, a dead silence; for the opposite is in fact the case. The embodiment of meaning in signs is not the beginning of a de-subjectification, or even de-validation. For writing is itself the key to precisely a new type of life, which means first and foremost the possibility of a different mode of activity with respect to meaning, or a different sense in which meaning can be taken up, or engaged in an interest. In fact, it is only because of writing that the horizon of a tradition is possible, that is, the horizon interests in which meaning does not appear as "purely original," but as a non-originary, *received meaning*—or the very starting point of Husserl's historical reflections, intended to reveal the possibility of the meaning of philosophy itself. All the activities of thinking, reflection, and articulating meaning that make up philosophy are oriented not around originary, but received meaning; the very attempt to clarify the theme of meaning itself, to fix a method with which to clarify meanings in light of originary intuitions, presupposes a prior possession of meaning as something unclear but nevertheless passively given, even understood.

The *Origin* thus represents a more developed analysis of what we have already described in previous chapters under the heading of "familiarity." The familiar is actual as a passive horizon of given meaning that associatively enriches whatever is originally, self-evidently given within experience.[30] But like any passivity, the familiar included, the embodiment of language is in the end of ambiguous value; it is both something on which one can depend, but it is also something which contains a risk with respect to the comprehension of sense. Especially in the case of writing, it is something on which a culture can become overly dependent:

> It is easy to see that even in [ordinary] human life, and first of all in every individual life from childhood up to maturity, the originally intuitive life which creates its originally self-evident structures through activities on the basis of sense-experience very quickly and in increasing measure falls victim to the *seduction of language*. Greater and greater segments of this life lapse into a kind of talking and reading that is dominated purely by association; and

> often enough, in respect to the validities arrived at in this way, it is disappointed by subsequent experience.[31]

The reliance on writing—and, by extension, on tradition as such—makes necessary a certain vigilance in order to preserve the integrity of thinking. Not because of writing itself, but because of the kind of understanding it makes possible, namely, passive understanding. Such an understanding works "associatively," which means that it draws connections that do not necessarily reflect what is present or given in original self-evidence. If we are led by what is possible in passive understanding, the danger is always present that we are being led away from clarity towards confusion. But this also means that the function of enrichment, the principal contribution of passivity, is itself a possible source of illusion. The enrichment brought about by passive life, if allowed to become a more and more dominant factor in the understanding of things, can seduce the I into accepting that which is only "given" passively, non-originally, thus merely associatively, as if it were, though not originally self-given, at least an acceptable substitute.

The problem of the reactivation of origins

However, such vigilance seems to be always too late; the acceptance of the non-original, but nevertheless "enriched" given is inevitably in place. Furthermore, there is no clean, identifiable break where the original has been left behind and the non-original taken up. The passively given and the actively given do not cancel each other out, but are two moments within any experience of original self-evidence. This is as true within the soul of the individual geometer as it is within the community. This means that vigilance against being confused by illusions that arise from an over indulgence in passive understanding is not enough; the point is not to avoid the passive, for in fact we are "originally" dependent on it. The passive makes possible the grip of meaning, to recall a theme from the Introduction. Instead, to invoke another theme from the Introduction, the point is to take responsibility for what is understood, by striving to relate it "back" to the truth of an original evidence of the self-given.[32] Only thus can one be "certain" that what has been understood is "grounded" in genuine truth, or in a genuine clarity with respect to givenness itself.

Yet the situation, and with that any act of responsibility involved or required, is actually more complicated, in a sense which directly involves the question of history. Historical life, or life in the form of a tradition, makes possible a particular mode of the experience of truth. It is not simply a question of tradition being available as a supplement to life, one source among others for the enrichment of the life of the present; tradition is rather the basis for a particular mode of possible clarity about the world, one that is genuinely historical in nature. And it is, as Husserl emphasizes, an experience of clarity possible only from within the use of a language. This clarity that history makes possible is not a pure, simple clarity, or the immediate givenness of the thing in its evidence, but is instead what can be called a *founded clarity*, though one that is "original" and "evident" in its own right, precisely because it has been *made* into an explicit, active, clear understanding. This explication is what Husserl in the *Origin* calls the "peculiar logical activity" of "clarification" (*Verdeutlichung*).

Take, for example, the experience of being passively led, through an association, or the directedness of a sign, to a meaning or signification. The result is that the meaning has been passively received, thus represents a "given" in this sense. It is not simply given as a beckoning, something to be thought about, but is already given in the form of an understanding, since it already comes with its own articulation, even if it is merely the directedness of a sign. Or more importantly, this articulation is experienced as such, as a kind of present thinking that one is aware of, but not "originally," for it is not something one has done oneself. A pattern of thinking needed to understand can be lived from within, but it can also be witnessed from without. But even if one remains on the outside, the very presence of this pattern presents certain possibilities for activity, even motivated possibilities—or, simply, interests. And by no means is the only interest, or even the most pressing, that of a reactivation, of accomplishing for oneself that which had originally given rise to what one is now aware of as "ready-made." Rather, the possibilities that are more likely to engage interest are those which have to do with possible *new* ways of understanding, even actively in original self-evidence, that are based upon what is passively given. All of which should remind us of Galileo:

> [...] it is something special, as we have said, to have the intention to explicate, to engage in the activity which articulates what has been read (or an interesting sentence from it), extracting one by one, in separation from what has been vaguely, passively received as a unity, the elements of meaning, thus bringing the total validity to active performance in a new way on the basis of the individual validities. What was a passive meaning-pattern has now become one constructed through active production. This activity, then, is a peculiar sort of self-evidence; the struc-

ture arising out of it is in the mode of having been originally produced.[33]

To be sure, for Husserl, original self-evidence already enjoys its own articulateness, which is then re-embodied in language, thus in a passivity that ensures its availability; but what this evidence is now available *for* is precisely an active understanding that realizes possibilities that have been opened only *given* this passively received, embodied meaning. Meaning is now more than meaning; passively embodied, it is open to being taken up in a kind of second life. We have already seen an example of this in the second life of traditional geometry in Galilean science; but there are also many examples from everyday life. For example, let us say that I describe to someone Caravaggio's "The Taking of Christ." Pursuing my description, the painting not only "comes back" to me in the form of a memory, or in imagination; both myself and the one to whom I am speaking "see" together the painting again, though now as a vision re-constituted and re-articulated in speech. In its re-articulation, the memory of the painting is now open to being "clarified," not because descriptions are always faithful to the original perception, but because a new kind of clarity can now be experienced, one which can take place only within language. It would be misleading to call this a "secondary" clarity, as if it were of limited importance; in fact, it is this kind of clarity that normally holds the center of gravity in everyday life, which more often than not values what can be said over what can be seen. Nevertheless, we still talk about the painting—*the* painting I *saw*, the Caravaggio that hangs in the museum. The painting as it appears in speech is not a new painting; the point is only that we can be perfectly clear and consistent about what we mean, without in turn subjecting this clarity to the standard of that upon which it is founded—the "actual," "originary" experience of standing before the painting. This, as Husserl points out, designates the domain of "logic" in the broadest sense: logic is the field of a self-contained *logos*, with its own particular mode of clarity, thus "truth." And it is a truth that is ultimately a constructed truth, which is in turn possible only given the passively acquired meaning formations characteristic of language.

This domain of *logos* is self-sufficient *in a sense*. As the accomplishment of activities that result in self-evidences of this "new type," life procures a line of understanding that can be pursued, and a world "understood," in a lively, forward moving manner that at first does not need to return to its original source or foundation. We can talk at length about the Caravaggio, finish the conversation and be satisfied that we have, once again, "seen" all that there is to see, precisely because we have said all there is to say—and we can do this without ever returning to the painting "in the flesh," for in fact it has received a new flesh in its second life within the concrete possibilities of a linguistic description. And so too with geometry: we can "do" geometry, perform constructions based upon the foundations already at hand; all of this in a lively, forward-moving manner which in turn forms the basis for new constructions, thus new "evidences," new "truths" based upon previous truths that have been handed down to us. And each new construction amounts to a *clarification*—a re-articulation of the same in a new form, which fixes it as something known. What is new is not necessarily always an improvement (Euclidean geometry is "just as true," just as apodeictic, as any space of *n*-dimensions in which the parallel hypothesis does not hold), but it is always a discovery of a different face of "objectivity" that had once been hidden, "undiscovered." All of this—the entire tradition of geometry as such—can occur without ever returning to the original self-evidence of mathematical ideality, i.e. to the original act of seeing that was "the origin of geometry."

Thus clarification, *Verdeutlichung*, has a passive basis. But if the point about the "seductiveness of language" is simply to be vigilant against the obscurity of language, or the arbitrariness of association that could lead one to the embarrassing situation of being caught thinking illusions, then there can be no question of a danger on this level. Science, the accomplishment of a life that seeks to clarify what is given, not only stands on solid ground, in the sense of truths that have been discovered in the past and to which geometers return, critically, again and again in verification, but it is actively pursuing the possibility of clarity as such, discovering ever new truths which are approachable *only* given the accomplishments of the past. All of this again points to the fact that science is as such a *success*. The interrelatedness of truths about the world—the complete, exhaustive explanation that is unfolding within the project of scientific thinking—has an unquestionable logical, rational consistency, and where this consistency is lacking, the movement of *Verdeutlichung* itself, guided by its own inner possibility, is already in pursuit of a solution. What is in effect or on the move in scientific life is a process of a radical self-clarification, which means, on the level of *Verdeutlichung* itself, a process of an unfolding articulation of an understanding of the world that seeks the ordered consistency of a system, and not only in its more philosophical moments. However, this systematicity of the world-articulation that is science is not simply an internal criterion having to do with the evidence or truth germane to the accomplished life of science, thus a life that rests (or can rest) on the passively/traditionally given. Part of its sense, Husserl argues, is that it also relates the original self-evidence of the world to the (other) original self-evidence that arises out of the constructing activity of science. And it is *this* relation, Husserl argues, that ultimately determines the genuine sense of "truth."

Let us consider this thesis more closely. If my articulation in speech of Caravaggio's "The Taking of Christ," as a constructed vision woven from concepts and observations, is a self-consistent whole of compatible assertions, building one upon the other in an ordered system (if it "makes sense"), then the tracing of the passively received basic elements of this description to the *original* self-evidence of seeing the painting "itself" will be the *ultimate* legitimation of my interpretation as "genuine," "true." This, Husserl says, is a *fundamental law*:

> Here the fundamental law, with unconditionally general self-evidence, is: if the premises can actually be reactivated back to the most original self-evidence, then their self-evident consequences can be also. Accordingly it appears that, beginning with the primal self-evidences, the original genuineness must propagate itself through the chain of logical inferences, no matter how long it is.[34]

Thus if everything in my articulation of Caravaggio's painting turns on the light emanating from the lantern that the self-depiction of Caravaggio is holding above the group of soldiers who are in the act of seizing the Christ, and if, upon viewing the painting again, I confirm that yes, indeed, the lantern illuminates the face of Caravaggio as well as the faces of the soldiers, then this confirmation is the origin of an authenticity with respect to what has been said in the interpretation. It "proves" it, we could say; but it does so only by bringing one certainty in relation to another, that is, the certainty of the understanding I achieved in the interpretation with the understanding that I achieve in seeing the painting "again," "in itself." The difference between these two certainties cannot be found in the general nature of certainty, if certainty were taken to be an homogeneous characteristic defining a class of assertions. The difference lies in the way that life, as a life of thinking within the passive horizon of a given, can turn certainty against itself, as well as use certainty to further justify itself. In this perspective, "truth" becomes a complex movement of articulation that can relate to itself on multiple levels within the constellations of "passive" and "active" that make up the structure of conscious life.

But suppose that I were unable to once again view the Caravaggio "itself," not even in memory, and that all I had at my disposal were a logically consistent set of concepts and articulations that I am no longer in a position to relate back to the original evidence, the original ground? Then the truth of this interpretation would be limited, its significance cut off at the source—that is, the source of its *ultimate* (but not *only*) legitimacy, though not necessarily its meaning or even its (limited) "truth," for it already has its own kind of self-evidence. Such would appear, Husserl argues, to be the case with geometry:

> The progress of deduction follows formal-logical self-evidence; but without the actually developed capacity for reactivating the original activities contained within its fundamental concepts, i.e., without the "what" and the "how" of its prescientific materials, geometry would be a tradition empty of meaning; and if we ourselves did not have this capacity, we could never even know whether geometry had or ever did have a genuine meaning, one that could really be "cashed in."[35]

If we were to lose the capacity to reactivate the very beginning, to again perform the act of discovery "for ourselves," instead of simply receiving the discovery as "given," then the tradition of science would be a non-genuine tradition, its evidence left uprooted due to the lack of the ability to even approach the evidence that was, and is, its "source." "Unfortunately," Husserl says in the next sentence, "this is our situation, and that of the whole modern age." Not only that. This principal condition for a genuine (justified) tradition *has never been fulfilled.* For the tradition to have remained "genuine" throughout, the capacity to re-activate the origin would have had to have been preserved *itself*, handed down as such; only then could the developing second life of geometrical self-evidence, the unfolding articulation of mathematical truths through the centuries, have taken the form of a "genuinely grounded" tradition. But that did not happen. And it was not an accident, an oversight, but a (fateful) omission that happened because of "easily understandable reasons," as Husserl puts it—just before he suggests that it could not have been otherwise:

> In the first oral cooperation of the beginning geometers, the need was understandably lacking for an exact fixing of descriptions of the prescientific primal material and of the ways in which, in relation to this material, geometrical idealities arose together with the first "axiomatic" propositions. Further, the logical superstructures did not yet rise so high that one could not return again and again to the original meaning. On the other hand, the possibility of the

> practical application of the derived laws, which was actually obvious in connection with the original developments, understandably led quickly, in the realm of praxis, to a habitually practiced method of using mathematics, if need be, to bring about useful things. This method could naturally be handed down even without the ability for original self-evidence.[36]

The issue here is nothing less than the formation of the motivation for "genuine" science, or science grounded in original evidence. This should not be confused with Husserl's conception of the telos of science. As a developing whole, science is directed towards its own completeness, or that final pattern of evidence which will fulfill the ultimate promise to fulfill the basic human need for a world. Yet this directedness is not by itself a motivation; the end does not pull us towards itself, at least not without our motivated activity. The point is critical: the genuine sense of science was not pursued by the original scientists, those who were scientists before science developed as a tradition; which means that what they accomplished, the discovery of ideality, finds its ultimate significance only in light of a concern that was not "in view" at the beginning. Instead, what was initially in view were concerns with other aspects of the discovery of ideality—practical applications perhaps being the most important, though one could also cite spiritual and religious consequences. Thus the meaninglessness of a science which no longer develops in an intimate connection with its origin, the "crisis," is at first an invisible meaninglessness, a hidden emptiness that itself must be discovered, and can be discovered only once it is motivated by the question of origin.[37]

The argument is that the need, thus the motive for grounding, as well as the entire problem of a "genuine" science, can arise only in the course of a tradition; it is a motivation possible only in the form of traditional being, or for a subject for whom tradition is a key element in the project of self-understanding. More specifically, it is only for such a subject that the task of taking on that unique mode of responsibility essential for genuine, fulfilled science is possible. Only such a subject is in the position to pursue the ultimate significance of science as a whole, because only such a subject can be open to articulating the idea of the "claim of science." Such a claim can be expressed only in the form of a tradition that makes a claim on "us," on those who are in the present, attempting to understand the possibilities of what could become through us, through those who are subjects within a historical horizon of the living present.

The origin of philosophy

The foregoing allows for a number of clarifications.

(1) First, what Husserl called the revealing-concealing genius of Galileo reflects the basic intentional-historical structure of the tradition itself. Galileo rediscovered the infinite, to be sure in a different form than the original discovery of geometrical ideality by the "first geometer," but in a comparable manner. Galilean science uncovers a new dimension for thinking, new possibilities for discovery and understanding, but in such a way that the significance of the discovery can and does, in the course of traditionalization, become *questionable*. Not in the sense of being challenged with respect to its veracity, but in the sense of being open to a question about meaning that had yet to be posed since Galileo. Again, this is not a story of an oversight; it is the history of an insight into the kind of questions that can motivate philosophical reflection, that can serve as its beginning. The argument is that such questions are possible only when they are addressed to an already received understanding, meaning, or significance—thus only when we open ourselves to the tradition, the past, from within the present, seeking to understand our place within the horizon of its possible significance. Thus a first thesis: part of the task of discovering the present task of philosophy is to ask the question of the place of the tradition in this present itself. "Our" place in a present will, to an extent, be determined by an understanding of the place of a tradition that we have made our own, in reflection (*Besinnung*), wherein we question the origin of its sense.

(2) In the *Origin*, one finds a more subtle conceptualization of the crisis thematic as a whole. Husserl's crisis is one of reason, of a rationality that no longer seems to order life, to give life the sense of itself necessary for its pursuit of itself, thus its future. But this rationality is nevertheless present "for us," it is a part of our world precisely as a tradition; the tradition of science is the presence of a successful project of rationality, thus it is a concrete presence that is a part of who and what we are. If this tradition no longer seems to have a place in our life, no longer seems to order human being in accordance with its claim to validity, then this indifference of the tradition to life provides the conditions under which the final and ultimate significance of the claim to truth embodied in science can finally be asked—and asked, according to Husserl, in genuine manner. That is, only now, precisely when reason appears to be empty, thus no longer "obviously" meaningful, can there be a true questioning in which the answer can be a "yes" or a "no." Sensitive to the crisis, it is no longer the question of simply making the claim, within the horizon of the tradition, as if the tradition itself were solid ground upon which we stand. Thus a second thesis:

the articulation of the "ground" itself is now possible, precisely because, within the crisis, it is no longer reasonable to simply assume that there is a ground.

(3) This question of the ground, of the meaning of the claim of science, is historical. For to ask the question about the meaning of science for life is not the same as to ask whether or not the claim embodied by traditional science is "true." That would presuppose that truth is meaningful, a standard by which the meaningfulness of science for life can be judged. To recall a suggestion from the Introduction, the real question of the crisis is whether or not the spiritual disappointment of the success of science has instead proven that the claim to truth is itself in the end meaningless, that it can no longer be simply accepted as the horizon within which the possibilities of what we are and can become are "obviously" projected. For science—more generally, reason itself, and with that a life in reason—is a formation of meanings motivated by the interest in truth, the truth of the world as a world of fact; in this way, science is itself an accomplishment of an understanding motivated by a particular idea. Science is historical in the sense of an ongoing formation of meanings in the wake of this original idea, or telos that defines the horizon of this particular interest of a life in truth, where all formations, both original and sedimented, form the body of the tradition;[38] *philosophy* is historical in the sense of posing the problem of reason from within the movement of historical life—that is, posing the problem of science in terms of its historical possibility as a claim formed in accordance with its idea. Thus a third thesis: the origin of philosophy is in the present, and it appears as the capacity for the present to question the past.

How is such a question posed? What type of problem does the philosopher engage for Husserl? We already have a clue: the regressive historical inquiry, the questioning of the tradition itself in terms of what is "implicit"—and what is meant here is precisely implicit possibility, an obscured sense that, upon reflection, reveals an unexpected aspect—in the significance of the tradition itself. The philosopher—or, better, the subject of the crisis, for whom the possibility of philosophy, thus of the present, is itself a question—questions whether there is an unexpected depth within the tradition that, upon articulation, could reactivate the sense of the tradition as a whole, precisely from within the question of what such a tradition *could mean* for life. Science, Husserl claims, no longer has anything to say to us; philosophy, however, is premised on the suspicion that, despite this failure, there may indeed be something to say. For the task of philosophy, in its questioning of the tradition within the question of meaning, is nothing other than *to speak to us*. Philosophy that does not speak to us, to who we are in all of our ambiguity, is dead chatter. Furthermore, the possibility of an implicit meaningfulness of the tradition is the possibility of an implicit meaningfulness of the present itself, in the form of the significance of the tradition for the present. In this way, the opening up of the historical problematic is itself an approach to the possibilities of life. This means that the activation of the "origin"—not in the form of an anonymous telos, but rather in the form of an articulated inner life of the tradition in self-evidence—is not the "recovery" of an act long forgotten, but is the "discovery" of what is possible *in principle*, a possibility that, for Husserl, must always be approached by an individual subject within the horizon of given tradition.

(4) There are three concepts in the *Crisis* that are tightly interwoven: *life*, *history*, and *world*. History, present within life in the form of traditionality, is the movement within which subjectivity confronts itself; and this self-encounter through history is not simply something that occurs in the world as one experience among others, but is inextricably fused with the very encounter with the world as such. What the world is, what it means, is in part decided in terms of this encounter between ourselves and our history. And what is possible for the historical present, what is possible with respect to the meaning of the world which is opened by life, or the "life-world," is determined by the essential possibilities inherent to historical becoming. If we can speak meaningfully about the apriori of the world, of possible world experience, this is so only because we can also speak of an historical apriori—the two are effectively the same, for in the form of subjectivity as life, all experience is historical. Yet how do we arrive at such an apriori? Is not the historical also the horizon of the non-given, thus that which cannot be grasped in terms of a givenness that would allow for reflection to reach that which is invariant, essential? This is where the argument for the givenness of history in the form of tradition becomes absolutely essential, and where the remarkable organization of the *Origin* begins to bear fruit: for at no place in this text does Husserl ever pose a problem or advance a thesis that passes outside the limits of his object—the presence of tradition in the living historical *present*. Because of this, Husserl can make the following claim:

> Through what method do we obtain a universal and also fixed a priori of the historical world which is always originally genuine? Whenever we consider it, we find ourselves with the self-evident capacity to reflect—to turn to the horizon and to penetrate it in an expository way. But we also have, and know that we have, the capacity of complete

> freedom to transform, in thought and phantasy, our human historical existence and what is there exposed as its life-world. And precisely in this activity of free variation, and in running through the conceivable possibilities for the life-world, there arises, with apodictic self-evidence, an essentially general set of elements going through all the variants; and of this we can convince ourselves with truly apodictic certainty.[39]

The *Origin* thus opens the way for a reflection within the lived present that will enable Husserl to ask the question of the apriori in such a way that its rootedness in life can become manifest. This rootedness itself, as we have seen in following Husserl's analyses of the origin of geometry, is itself historical in nature. But what such an apriori could be for Husserl—its character, status, and availability for reflection—is not a question that can be pursued from within the *Origin*, but only by turning to the theme of the lifeworld as it is developed in the text of the *Crisis* itself, which is the task of the next chapter.

Notes

[1] The title is not Husserl's, but Fink's, who published this working manuscript of Husserl's, which dates from 1936, under the title "Der Ursprung der Geometrie als intentional-historisches Problem" in the *Revue internationale de philosophie* in 1939.

[2] "Warum bedarf der Philosoph der Philosophiegeschichte?" Hua VI 495:28-29 (Beilage XXV, from 1934).

[3] Crisis 353-354; Hua VI 365:17-31.

[4] For an elegant presentation of Husserl's conception of idealization, using as an example Galileo's analysis of free fall from the *Discorsi*, see Garrison "Husserl, Galileo, and the Processes of Idealization" (1986). Cf. the discussion of Galileo in Gaukroger, *Explanatory Structures* (1978).

[5] Cf. EJ/EU §65. But can one say that, for Husserl, the phenomenality of ideality is repetition—and if so, does this not come close, as Deleuze would argue, to equating *repetition* with *universality*? (Cf. Gilles Deleuze, *Différence et répétition* (1968) "Avant-propos.") If so, it should be recognized that it is also the result of a key self-criticism of another questionable equivalence, that of the identity of meaning (*Bedeutung*) with species identity in the I. Logical Investigation (LI,1/Hua XIX,1 §§31-35). There Husserl argues that meaning as a species unity is instantiated in various "moments" of individual, real mental acts of meaning. Husserl later rejects this and argues instead (cf. EJ/EU §64 a-b, d; LII 13 and M III 3 IV) for a strict identity of judgments across individual acts of judging—an ideality of pure repetition, as opposed to the species ideality that would carry with it the traditional conception of a bond, as well as a difference, between species and individuated moments of the species. Cf. the discussion by G. Heffernan, *Bedeutung und Evidenz* (1983) 83-106.

[6] Cf. EJ/EU §64c.

[7] J. Derrida, in his *Edmund Husserl's Origin of Geometry* (1968, 1989), makes two observations in this respect. First, on p. 70, Derrida argues that the German word "Löwe" is the ideality which appears only within a "facto-historical language", i.e., German as a historical linguistic community, thus manifest within space and time. However indifferent it may be to its sensible manifestations in writing and spoken exemplars, it remains, as a "German word for...," bound to another complex of the empirical (though empirical qua, Husserl would say, spiritual objectivity). Second, in footnote 62 on page 67, Derrida refers to a similar presentation in Hegel of what we have been calling the indifference of ideality in language, citing in particular §462 of the *Encyclopaedia*, which falls within the discussion of memory that McCumber has argued is central to understand Hegel's conception of thinking. See John McCumber, *The Company of Words: Hegel, Language, and Systemaitic Philosophy* (1993); also Maria Leao, *Art, Language, and Thought* (1996).
[8] Crisis 357; Hua VI 368:30-39.
[9] Cf. EJ 267; EU 321:7-13.
[10] Thus in the I. Logical Investigation, where Husserl poses the question what expressions "mean," the answer is not the meaning of the expression as an empty, unfulfilled "Meinen," but precisely the fulfilled meaning, or *intentio*. See LI,2/Hua XIX,1 §34.
[11] Ideas I/Hua III,1 §88.
[12] Nevertheless, even free ideality is bound, insofar as it is objective—that is, it is bound to the condition of its own objectivity: being discovered. Cf. EJ 267; EU 321:28-30.
[13] It should be noted that the theme of "noema" is not limited to phenomenology. Husserl himself cites the Stoic doctrine of *lekton* at FTL 82; Hua XVII 86:39-87:6. Cf. Sextus Empiricus, Adversus Mathematicos VIII 11, 38. Also see J. M. Bochenski, *Formale Logik* (1956) 126-127. What Husserl finds interesting in this doctrine is, presumably, both the subjective focus as well as the denial that this subjectivity is in and of itself manifest in the form of a *psychic corporeity*. "*Lekton*" means: what is meant (*sêmainomenon*), which is for the Stoics clearly a product of a logical act of presentation (*phantasian logikên*), but which is at the same time non-corporeal, unlike the sensible formation in which the meant is meant (*semainon*) and the thing itself (*tugchanon*) The "object" of Stoic logic, therefore, even if it is the product of thinking, is not thereby something subjective as opposed to something objective, but an *ideality* opposed to both.
[14] Crisis 359; Hua VI 370:1-12.
[15] For different ways that this tension can be spelled out, cf. in particular Ideas I/Hua III,1 §§91, 99, 131-132.
[16] See André de Muralt, *L'idee de la phénoménologie* (1958) §22.
[17] De Muralt, *L'idee de la phénoménologie* (1958) 116: "C'est ici le langage qui témoigne de l'objectivité idéale de notre connaissance et nous permet de la motiver."
[18] Nor is it a problem that Husserl himself ever fully resolves. He neither takes the initial step of making clear what would motivate a "phenomenology of language," nor does he himself pursue such a project. What would such a project resemble? If the key is to look upon language as an acquisition that we possess in the form of an ideality, but an ideality which, so to speak, encompasses the whole of the world (everything has a name; everything can be expressed, even ideality itself in its objectivity), then would not a "philosophy of language" in the spirit of Husserlian phenomenology have to resemble, in some important respects, what the German Idealists and Romantics used to call the "Philosophy of Art"? Schelling in particular, for whom language itself was a perfect "work of art," and for whom the task of the philosophy of art was to present, in thought, the Absolute that is present, in "reality," in the artwork itself. And the key insight of Schelling's work, that the manner in

which art is a manifestation of the absolute is by virtue of its infinite interpretability, an interpretability which is continually in motion even as the work itself remains indifferently the "same," is perhaps an insight that needs to be pursued in understanding the ideality of a language that need never be understood as a hindrance to the unfolding movement of manifest content/meaning. See Schelling's *Philosophie der Kunst.* In: *Schelling's Werke*, Bd. III (1927).

[19] "Naturally, we shall not go into the general problem which also arises here of the origin of language in its ideal existence and its existence in the real world grounded in utterance and documentation [...]", Crisis 358; Hua VI 369:11-14.

[20] A basic difference that divides Husserl and Heidegger can be discerned in light of this. For Heidegger, the past, in the form of tradition, is something that is authentically appropriated from out of the movement of our future, thus precisely where tradition as the presence of the past does not guide the sense of the investigation. As beings who are in essence futural, we *confront* our past ("tradition") in terms of the possibilties opened by the temporal extases of our Dasein, possibilities which are at stake at any given historical moment. Cf. *Being and Time* (1929) §6. For Husserl, as we can see in the *Origin*, it is almost the opposite: the future is something that is already in place, already in effect, and which must be discovered in and guided by the past. The place of our future in our lives, the aspects of our current situation that allow for a new development, a new understanding, whatever there is that allows for a way to address the crisis of the sciences—all of this is for Husserl prepared by a past that is by no means a static order, but always already *on the move.* This difference between the two founders of phenomenology is extremely important, so much so that I cannot pursue an analysis here and at the same time do it justice. For my purposes, I will leave open the question whether the Heideggerean challenge deals a fatal blow to Husserl's program, or whether on the contrary leads to just the opposite.

[21] "And yet there lies in this lack of knowledge [*Unwissen*], everywhere and essentially, an implicit knowledge, which can thus also be made explicit, a knowledge of unassailable self-evidence. It begins with superficial commonplaces, such as: that everything traditional has arisen out of human activity, that accordingly past men and human civilizations existed, and among them their first inventors, who shaped the new out of materials at hand, whether raw or already spiritually shaped. From the superficial, however, one is led into the depths. Tradition is open in this general way to continued inquiry [...]" Crisis 355; Hua VI 366:28-36.

[22] Crisis 355; Hua VI 367:4-7.

[23] Crisis 356; Hua VI 367:24-32.

[24] Crisis 359-360; Hua VI 370:25-36.

[25] Crisis 360; Hua VI 370:37-45.

[26] Crisis 360; Hua VI 371:6-14.

[27] At this point, before taking up the question of writing, which plays such a conspicuous role in the *Origin*, one might be interested in recalling the idea of *genuine teaching* that Plato's Socrates presents, albeit in poetical language, in the *Phaedrus*. Here teaching is not primarily speaking a language someone understands in order to impart information, but is instead an activity meant to aid in the formation, and to above all *witness* the formation, of those aspects and features of the soul of the beloved which, *in concreto*, are a manifestation of the idea, and which are thus objects of enjoyment for the teacher, who is the "lover" of the idea. The true moment of learning for the student is not, however, attained by merely being that which reflects the idea, thus in being admired by the lover—instead it occurs precisely when, in seeing his own reflection in the eyes of his lover/teacher, he discovers in himself the possibility of loving something which is not itself concrete, which is not himself

or of himself, or of the world at all, but which is precisely ideal. In this moment of discovery, an interest, or eros, awakens in which the beloved, in a process beyond language and speech, beyond communication, becomes himself a "lover." See *Phaedrus*, 255c-e.

[28] Crisis 360; Hua VI 371:18-25.

[29] Cf. the footnote on Crisis 361; Hua VI 372, where Husserl points out that the reactivation of a linguistically embodied meaning "is by no means necessary or even factually normal. Even without this he [the reader] can understand; he can concur 'as a matter of course' in the validity of what is understood without any activity of his own. In this case he comports himself purely passively and receptively."

[30] This is a familiarity which even enriches, or gives shape to, what is unfamiliar. In the middle of this sentence someone interrupted my work by calling my number, but could not speak English. The unfamiliar words had, nevertheless, an echo with words from languages with which I am familiar; the sound of his voice, though I had never heard precisely this voice before, had consonance with the multiplicity of voices that occupy my life day in and day out—all of this is at work even before I make a judgment that "this is a man, he is speaking Spanish, he is interrupting my work." All of this is at work, but not complete, for only the judgment, the act in which evidence comes into its own as the *Selbstgebung* of an "object," ultimately "forms" the event in its full meaningfulness; still, the passive horizon gives shape, tone, and color to the experience—and is for Husserl essential for understanding the character of experience as a *lived* meaningfulness.

[31] Crisis 362; Hua VI 372:17-25.

[32] In connection with this, let us quote Husserl's footnote at Crisis 362-363; Hua VI 373: "At first, of course, it is a matter of a firm direction of the will, which the scientist establishes in himself, aimed at the certain capacity for reactivation. If the goal of reactivatability can be only relatively fulfilled, then the claim which stems from the consciousness of being able to acquire something also has its relativity; and this relativity also makes itself noticeable and is driven out. Ultimately objective, absolutely firm knowledge of truth is an infinite idea." The conditions of the discovery of this idea as something we are motivated to take up is a topic that can only be touched on here; its full analysis would require a detailed presentation of the genetic phenomenology developed by Husserl in the 1920's.

[33] Crisis 364; Hua VI 374:23-31.

[34] Crisis 365; Hua VI 375:15-20. Here again the issue of truth as an infinite idea makes itself felt, particularly with the case of geometry and science: "However, if we consider the obvious finitude of the individual and even the social capacity to transform the logical chains of centuries, truly in the unity of one accomplishment, into originally genuine chains of self-evidence, we notice that the law contains within itself an idealization: namely, the removal of limits from our capacity, in a certain sense its infinitization." With this description one now has a better sense of the conditions for the possibility of genuine truth as an interest for thinking (thus a philosophical interest)—the infinitization of man, as was already pointed out above in the discussion of Galileo.

[35] Crisis 366; Hua VI 376:10-16.

[36] Crisis 368; Hua VI 377:31-44.

[37] Crisis 368; Hua VI 378:4-7: "Thus also it is understandable that the lost original truth-meaning made itself felt so little, indeed, that the need for the corresponding regressive inquiry had to be reawakened. More than this: the true sense of such an inquiry had to be discovered."

[38] "We can also say now that history is from the start nothing other than the vital movement of the coexistence and the interweaving of original formations and sedimentations of meaning." Crisis 371; Hua VI 380:19-21.
[39] Crisis 374-375; Hua VI 383:11-23.

Chapter Five

The Problem of the Lifeworld (*Crisis* §§28-34)

> *Will man denn die Wissenschaft noch >>lebensnäher<< haben? Ich denke, sie ist schon so nahe, daß sie uns erdrückt. Eher brauchen wir die rechte Lebensferne, um noch einmal einen Abstand zu erlangen, in dem wir ermessen, was mit uns Menschen vor sich geht.*
>
> —Martin Heidegger, *Die Frage nach dem Ding* (1935/1936)

The givenness of the world

What is Husserl's concept of the lifeworld? It is often described in somewhat vague terms as the "world of life," the world of our "everyday" dealings with things. But it should be stressed from the start that the life-world in Husserl's writings is a carefully developed technical concept and not simply a reference to some common sense notion of the world. It is a concept developed for a specific use, and carries connotations that would not normally be in play when speaking about familiar forms of everyday life. It is important to be sensitive to this. The function of the concept of the lifeworld is to highlight a tension that germinates deep within the familiarity with the world, one that is intimately connected to the developing question of the meaning of science as it is experienced in crisis. This tension is of the utmost importance for Husserl, since its analysis will in turn provide a guiding clue to the transcendental problematic.

What is this tension? Two things should be emphasized from the start. (1) Science itself, insofar as it is present in the form of a tradition, is a part of the lifeworld; it has its place, precisely as something with which we are familiar. Science is a familiar, habitual manner of understanding, one that is just as "everyday" as any other mode of thinking. But at the same time, science also stands in tension with the lifeworld; which means, among other things, that it is in tension with its own place in the fabric of the familiarity with things. This tension was already something prominent above in the discussion of Husserl's concept of crisis. Science belongs to the world, it is familiar; but it is also that through which this world has become strange, a meaningfulness that has somehow at the same time become empty of meaning.[1]

Further, (2) even in the practice of science, the theme of the life-world is partially developed, for science is by no means oblivious to its own situatedness. Scientists have no need for philosophers to tell them that they are subjects, that they live in a world shaped by subjective interests; on the contrary, the subject-relative is of utmost concern in methodological attempts to exclude, or at least mitigate, all of that which obscures the goal towards which science strives, the objectivist ideal of a clean articulation of the world of facts. This means that, already within science itself, the life-world is no longer anonymous, but already a distinct theme; though it becomes a theme in a distorted fashion which, paradoxically, secures its anonymity by insisting that it is of no value. Thus if science is a distortion for Husserl, it is not because it provides a false picture of the way the world is "really experienced"; it is a distortion due to the naiveté characteristic of the objectivist position: the position which holds that the subject-relative can simply be excluded from our considerations of the truth of the world, that a pure articulation of facts by a radically self-effacing subject exhausts the possible shapes of the rational.

This is a distortion, because it suggests that the tension between "world" and "subject" has nothing to do with the positive meaning of the world. And it is not enough simply to point out to the objectivist that even science itself is "subject-relative," namely, dependent upon human subjects for its very existence and value. Such a point could be easily conceded, for example by an objectivist who chooses to read such subject-relativity in terms of a potential limitation of the possibility of full, complete knowledge of factual reality. The result is that the objectivist could be as convinced of the irremovability of the subject-relative as any transcendentalist, arguing that it is the fate of finite subjectivity to never fully transcend itself, to necessarily remain hampered by the distance between finite, limited subjectivity and the infinite expanse of the horizon of the factual world. But that the difference between subject and world, in another respect, is instead constitutive; that the distance marking out the encounter of thought and being forms the context within which the very sense of objectivity is possible at all; that the subject-relative, far from being the diminishing of objectivity, is the very condition for the possibility of the manifestation of the sense of the objective as a positive human experience—all this is hidden from the objectivist.[2]

If one begins with this theme of tension, along with the naiveté of objectivism, then it can be shown that the lifeworld in Husserl's reflections is defined, or becomes definite, only in contrast to the kind of clarity sought by objectivist science. For the lifeworld represents what I would like to describe as a certain resistance, or reserve *vis-à-vis* what is accomplished in and as scientific clarity.[3] The truth of the world, of things, is always "more" than the truth of science—but this does not devalue the latter, for neither lifeworld nor scientific clarity has an exclusive claim to the description "true." Instead, the lifeworld, one could say, is the truth of the world before this truth has become clear in science. This rather awkward way of putting the matter is necessary, though it can be misleading, in that it could easily be interpreted purely objectivistically, where the accomplishment of objective science is simply being recognized as a becoming-aware of the "in-itself" of the world which is already "there," waiting to be investigated by a theoretical subject. This awkwardness, nevertheless, is indicative of something that could turn out to be quite important: however distinct the "lifeworld" and the "world of objective science" may turn out to be, when looked at another way there is no distinction at all—for there is an important sense in which the world as it is arrived at by theoretical understanding is simply the same world as the world of life, only now "understood," "clarified" in a more "adequate" fashion.[4] To better understand this distinction, several preliminary considerations are necessary for an anticipatory intuition of something that can be developed in detail only once we have entered into phenomenology proper.

(1) First consideration: one could ask, perhaps with a certain naiveté, why it is that we must speak of the "lifeworld" qua "world" at all, and not simply accept that the world is the way that science describes it, even for everyday life. Is not the lifeworld simply the world explained by science? However much it can be said that the ingrained habits and experiences of pre-theoretical life may pre-empt what we "truly" understand in the scientific attitude, once this attitude has been adopted, does it not simply take over and expand on whatever knowledge natural life had been able to accumulate before the beginning of scientific discovery?

The key to an answer lies in the concept of "world." This is a theme that has been used already extensively above, but it needs to be introduced again, though with a somewhat different emphasis. Namely, it needs to be stressed that the world is not simply that which is understood, but that the world is that which is *given* in understanding. And if one continues to think of understanding as a kind of articulation, thus as a kind of "accomplishment," then the world is not only that "about which" we speak when we conceive, think, act, and reflect, but it is also characteristic of a givenness within which we move, as those who are able to navigate this givenness by relying on an understanding to provide an orientation. Because this givenness of things, of the world of things, is open to being conceived, our encounter with things occurs within the horizon of an understanding that, over time, grasps things in the form of "having been understood to be such..." The world is in this way the development of a "view" of things; but it is a view that always presupposes the givenness of the world of things as that "about which" we speak, think, and with "respect to which" we act. In short: it is not simply the "world" we understand and experience, but its *appearing*, where appearance itself should never be simply identified with concepts and understanding in general.

If so, then it is not the case that understanding projects a structure onto a mute world; the world itself, as that which is given, sets into motion an encounter that, however much it may be anticipated, is always "in action" just before our thinking response, our understanding. What was above described in terms of the accomplishment of understanding in the formation of the surrounding world (*Umwelt*), where the understanding responds to a need for a world, is something which nevertheless occurs "in" a world that is already in place. The world is always that "in which" our understanding takes place, even if it is also that "about which" our understanding understands, and in doing so provides access. Our grasp of things, our access, is always "for us" preceded by the things we grasp.[5]

This givenness (or rather pre-givenness) of things plays a complicated role in the kinds of meaning that things can have, as can be seen from Husserl's genetic analyses that can be found in his lectures from the 1920's. For my purposes here, all that I would like to stress is the thesis that the world is pre-given in the mode of something that, because it is open to being articulated, thus to being meaningful, represents a basic foundation for the formation of motivations that structure the interests of the subject who understands. Something I do not understand, something with which I am unfamiliar, is nevertheless "there," "given" within a horizon of possible movements of understanding, possibilities of which I may be only dimly aware, perhaps momentarily fixated precisely by the strangeness of the given against the backdrop of the familiar. And if I am to be able to deal with this unfamiliar something, I must discover its secret—the key to its identity, which is identical to the key to its features, characteristics, behavior, and style. The necessity for such an articulation is an important point. The world is open to being articulated; which also means that it is not given as something the meaning of which is already in place. To have the world does not mean to have also acquired its sense; meaning is always something arrived at, which means that it is first of all pursued, lost, gained and re-gained in different forms. One of Husserl's basic theses is that the world is open for a meaning that can be acquired only thanks to the activities and accomplishments of subjective life; it itself does not simply "give" a meaningfulness that would always already be assembled for a subject by the simple virtue of its being present. Thus the only way to deal with things is to pursue them within the horizon of their possible articulations, even in those cases where one of these possibilities is actually a "well-established" way of thinking and acting. Even established, well-worn thoughts and truths must be *thought*.

The world in its pregivenness, thus that which motivates its own acquisition as articulated meaning, is for Husserl the most primary, and therefore original mode of "having" a world. It is a level of world-having in which the experiencing subject is not yet in possession of its truth, or has not yet arrived at the final form of knowing that would represent a full grasp of those validities that express its true being. The idea is that the very givenness of the world, prior knowledge, already takes the form of a "having," where even if the need for a world has not yet been satisfied, its satisfaction is nevertheless prefigured by a mode of proto-possession—instead of knowledge, the subject has a *doxa*. Thus for Husserl the original, first acquisition of the world, the first encounter with the world as something open to being meaningful, occurs within the accomplishments of encountering the world of things in doxic modalities that represent a "grasp" of the givenness of the world in its openness to its further articulation, in the sense in which a belief is open to being verified, validated, clarified. But also: in the sense in which a belief is a knowledge in need of being knowledge, thus, paradoxically, in need of itself. The result is that the need for a world is embodied within the forward moving development of knowledge as such, where it is from the start determined in the form of a *need for knowledge*.

Lifeworld for Husserl is the world of *doxa*, of "belief," for it is within doxic life that we initially have the world as a ground for further articulation, further thinking and fixing with concepts. And, with respect to the theme of "givenness" itself, it is important to point out that the world is no more "given" when fixed in concepts than it is in its unfixed, ambiguous form on the level of doxa. Science questions, and quite rightly, belief as a mode of legitimate understanding, or a form of successful grasping of the world in its meaning; but it can never supersede the doxic lifeworld with respect to the ultimate having of the world as such, but on the contrary presupposes this originary accomplishment and acquisition of the lifeworld as its *ground*.[6]

The question of the world

Science is an activity, and accomplishment, of an understanding that operates in the space opened by language and the givenness of the world in doxic life. Motivated by what is already accomplished in this life, science enriches the meaning of the world encountered in its givenness; it is a grasp in clarity of "the" world in which we live. But this accomplishment of science, which is a response to the need for an articulated world, nevertheless develops only in tension with the world. This also means that it develops in tension with the very horizon within which it develops. This is because, however successful science may be, the horizon of its development never loses its doxic character; however established its guiding idea, the accomplishment of science, as having "arrived at" the truth of the world, never ceases to bear the mark of a certain relativity. To "know" and "understand," to arrive at a validated truth, is to be bound in turn to a particular horizon of "belief."[7] This makes sense only if one does not continue to think of belief as incomplete knowledge, and nothing more. To fail to do so, is to become entangled in empty expectations that the success of science will replace belief, as opposed to enriching it. Instead, the argument of Part IIIA of the *Crisis* is that belief is the foundational achievement; *doxa* is, in a sense, complete in itself; we do not need science in order to have the world "as such," but we do need to "believe." It is only belief in the world that can be taken for granted as the horizon within which all other thought and understanding takes place; where thought and understanding function as enrichments of the original having of the world through belief.

Another way to express this thesis of the *Crisis* is that the validities or truths achieved by the tradition of science fill out belief, giving it a more varied and profound inwardness, which in turn fulfills belief as an experience of what belief has always already been in the possession of—a "world." Belief, however accepted and complete, is also, paradoxically in need of itself; this is what science provides. If so, then even in science validity does not really replace belief. This also excludes the possibility that the ideal of knowledge as the consciousness of a truth that could not be otherwise—or even if it "could" be otherwise, it "is" not—could, as an accomplished cultural project, take over the world-giving function of life as a life of belief. Instead, scientific knowledge—or any understanding which rests on tradition, culture, human accomplishments—occupies an original space of world-belief, without being identifiable as this space itself. Even when science takes over as the dominant style of cultural understanding, making its claim about what is meaningful and meaningless, there is nevertheless a fundamental irremovability of the doxic which, however obscured it may become (and perhaps always is), plays a constitutive role with respect to the situatedness of the meaning of whatever validities form the framework of understanding. The doxic remains, to recall the theme of obviousness in the discussion of Galileo above, formative of the confines of the very situatedness of science itself. The concept of the lifeworld is a first step towards understanding this irremovability of the doxic. More, it is a first step towards understanding how it is that the doxic is not only symptomatic of some sort of "epistemological limitation" of a cognitive subject, but is rather itself the ground upon which rests the very possibility of science as a cultural achievement.

Let us take a closer look at this thesis by coming at the issue from another angle. What "is" the world for science, originally? That is, the question is not about what science claims, with whatever level of justification, the world to be, what its "nature" or "structure" is. Rather, from the perspective of science as an activity, as an accomplishing articulation, what is the most original form that the world takes, in order precisely to motivate the activity of science? One could then ask, recalling the discussion above of the nature of the crisis: is it not plausible to say that, from the perspective of this activity as an *experience*, the original givenness of the world for science (and perhaps all cultural activity) is essentially *problematic*? To be sure, questions about the world are always framed not only in terms of the given in immediate experience but in terms of given concepts and expectations; questions do not arise from the world ready-made, we must learn how to question. But if the givenness of the world is such that it makes itself available to an understanding by motivating its articulation on the level of sense, of givenness, then is it not the case that it does this precisely by giving itself in the form of a question, to be both formulated and addressed?

Following this suggestion, the problem of the lifeworld can be framed in the following way. As the original acquisition of the "world," the lifeworld is in fact the achievement of a fundamental understanding of the meaning of the world, a meaning in which the world is unmistakably given.[8] But what is thereby acquired is the givenness of the world as a question, where the givenness of things is grasped in the form of the horizon of the questionable. Other, "higher order" acquisitions, such as science, are "founded" on the lifeworld in the sense that they are motivated activities that address the given from the horizon of its questionability; more, these activities are meaningful only as responses to the questionability of the given. Science is, in its idea and cultural reality, an "answer," which above all means that it is at least in part "founded" on the question it is meant to address. To emphasize: it is *in the question* that the *original* sense of the meaning of the world is achieved, *not in the answer*, which itself is something that can only enrich this original understanding of *what is being asked*. The thesis that the world of science fails to supersede that of the lifeworld can be read to imply that any "answer" or "response" remains in tension with the question it "intends" to address. In short, the thesis is that the questionability of the world is irremovable.[9]

Looked at from this perspective, the lifeworld should not be understood as an acquisition in the sense of another, alternative world with its own logic, its own rationality. What is acquired in the world as question is the sense or meaning of the world as a beginning, a starting point for a task of understanding. The fact that it is irremovable means that it is a beginning that never ceases to be a beginning, and thus is never really in view as something to which we can "look back"; yet it is nevertheless "understood." This aspect of the world becomes apparent, or itself an explicit issue, only when the problem of the beginning, not in time but in meaning, has itself eventually become necessary, as is the case, Husserl argues, in modernity; only when we are faced with the task of relating our understanding of the world to a primordial beginning as ground of possibility does this reticence of the questionability of the world in the face of our understanding become a philosophical theme. The "lifeworld," in this respect, is a uniquely philosophical theme, designed in light of the motivations implicit in the project of reflection as *Besinnung*.

Modality, problematicity, apriori

Such an interpretation helps make sense of several features of Husserl's presentation of the concept of the lifeworld in the *Crisis*.

(1) The first is the interest in modalities; or, better, in modal phenomena. In Husserl's writings, even before the *Crisis*, the topic of modality is not limited to the study of the truth-values or semantics of modal assertions or propositions. Rather, for Husserl, the richness of givenness is a direct consequence of its modal structure.[10] It is because of this structure that the given is open to its own interpretation in the unfolding of an experience, as well as to playing different roles in the constitution of validities (*Geltungen*). An important example of this modal structure for Husserl is a pre-predicative, pre-cognitive form of *negation*.[11] It is important because negation on the pre-predicative level can be understood as a type of preservation: it is not a canceling out of givenness as such, but itself a form of belief, or a possible contortion that a belief can take with respect to its function. Namely, a belief that has been negated does not disappear, but

"remains" in consciousness, and its presence as negated—or its negated presence—opens the way for the realization of other possibilities of belief, possibilities that receive their support and motivational density precisely from the presence of the negated belief. This preservation of negation is operative even on the level of predicative thought, which is an example of a doxic structure basic to higher order cognitive orders. Husserl's favorite example is mistaking a wax mannequin for a flesh and blood person: when I discover that what I had took to be an exotic woman is really only a cleverly sculpted wax figure, the "negated" judgment is not simply replaced by the judgment "this is a mannequin," but remains as a functioning component of the intentional experience, thus part of the development of the meaning of the given that is given in this experience. Modality is in fact the condition for the possibility of experiences of this type, that is, experiences in which meanings depend on the contrasts and overlappings that can be manifest only given the passive syntheses that lie at the heart of modal phenomena. Other examples include the suspension characteristic of doubt, the verification of certainty, and so on—the important point being that the modalities of the given form the basic, foundational structures of whatever shape the world takes for understanding.[12]

(2) This interpretation also bears on the important theme of evidence (*Evidenz*). The lifeworld is for Husserl the region of "original" evidences—but this does not mean that the lifeworld is a region of ultimate, simple clarities, or of final ("adequate") knowledge. Evidence in Husserl is not equated with a knowledge that unfolds in accordance with the standard of clarity and distinctness. For Husserl, the ideal of clarity and distinctness can be fulfilled only in a re-formation, or re-acquisition of evidence, a repetition that does not itself replace what had already been achieved in original evidence—which is precisely the givenness of the thing "itself" (*Selbst*), whether clearly and distinctly or not. A passage from *Experience and Judgment* is characteristic of this distinction:

> As "*self-evident*" [*evident*], then, we designate *consciousness of any kind which is characterized relative to its object as self-giving this object in itself*, without asking whether this self-giving [*Selbstgebung*] is adequate or not. By this, we deviate from the customary use of the term "self-evidence," which as a rule is employed in cases which, rigorously described, are those of adequate givenness, on the one hand, and of apodictic insight, on the other. This mode of givenness, too, is to be characterized as self-giving, i.e., of idealities and of general truths, but

> every kind of object has its own mode of self-giving, i.e., self-evidence, even though apodictic self-evidence is not possible for every kind, e.g., not for the spatiotemporal objects of external perception.[13]

Thus if the lifeworld is the sphere of original evidence, this does not mean that it is the sphere in which understanding is the most "genuine," untainted by false ideas, whether those of science or other cultural constructions. The evidence of the lifeworld is rather the emergence of the original self of things, along with the kind of understanding this emergence entails; it is also the place of the original self of the world, that self which the world is originally given "as." And this "self" must be distinguished from what is arrived at once the justifications of clarity become the goal, for clarity is something into which the original evidence, or *Selbstgebung* of things must first be *brought*. In connection with the discussion of the *Origin* above, it is clear that this "questioning back" to the lifeworld is not the search for an originary clarity that would dispel the fog of the false ideas of tradition; for it is only in the activity and accomplishment of tradition that justified, established knowledge and clarity is possible at all. The point is rather to bring the clarity achieved by the tradition back to the original "evidence," the primal "self" of the world, in order for this self to be *genuinely* brought forth into the already accomplished clarity of knowledge represented by science. Husserl's argument is that the originality of world-evidence makes knowledge possible, without thereby becoming identified with it, as if in the establishment of the latter the original sense of the world would be replaced. More, this reserve of the lifeworld is built into understanding as such—for the lifeworld is, after all, the world that we understand; within this grasp that we have of things, there remains a sensitivity to this difference, or to the tension between knowing and known, what is understood and the understanding of it. Even in the wake of the formation of knowledge, and what we see thanks to our capacity to know, there remains this sense of the world as *pre-given*.

This fundamental acceptance of the world as a pre-given horizon of activity and understanding is not blind naiveté, but the grip of an understanding sensitive to this tension between knowing and known. The "givenness" or "original evidence" of the world is something that can be taken for granted, precisely because this givenness "as such" is the origin of all questions pursued in experience. Thus it is in this sense that the world itself originally takes the form of a kind of question: it is a question that is always already in place, already an "issue," which means that there is something paradoxical about turning around and putting the world in question; though it is not absurd. To question the world, itself already a question, along the lines of Leibniz' "why is there something and not nothing?" would be equivalent to the attempt to think the beginning of the possibility of questioning itself. Thought of in this way, the alternative to the world, that there is "something," would not be "nothing," but rather a kind of silence that would "give" nothing, motivate nothing, demand no answer.[14] The question would perhaps be: why does the world demand from us an answer, why does it make itself an issue, and not remain indifferent to being understood?

This also implies that the "naiveté" of what Husserl calls the natural attitude is not the absence of questioning, an acceptance that would have nothing to do with the experience of problematicity. On the contrary, naiveté on this reading represents the acceptance of the question of the world, thus a horizon in which the meaning or sense of the world is something pursued—a horizon opened, and possible at all, only given the givenness of this question. And it is a horizon pre-given, accepted by a life that has, in its understanding, always already engaged the possibilities of the problematicity of the world "itself."

(3) This interpretation also bears on the difficult concept of the "apriori of the lifeworld" and its relation to the idea of an "historical apriori" that one encounters in the *Origin*. The question here, I would like to suggest, again turns around the thematization of a tension, this time within the apriori itself. As in the case of the distinction between the world as it is understood by science and the world as it is "lived," there is within the apriori a productive tension that enriches the content and structure of the whole. Though in the case with the apriori there is an important difference, one that draws from a basic thesis of phenomenological philosophy since the *Logical Investigations*. As Heidegger stresses in §7 of his 1925 lecture course *History of the Concept of Time*, what is distinctive about the conception of the apriori in Husserl's phenomenology is the implication that the apriori is indifferent to subjectivity. This stands in contrast with the Cartesian/Kantian tradition where the apriori is identified as a moment within subjective comportment, where the apriori is equated with certain structures of this comportment that are in place "prior to" what is given in those experiences which these comportments have in turn made possible. Thus if knowledge is a subjective comportment, then what can be known about the accomplishment of knowledge "in itself" counts as a kind of apriori; this can in turn be interpreted as an inner knowledge of subjectivity, and if what can be established as belonging to this inner sphere of subjectivity includes synthetic judgments apriori, then we can claim to have arrived at apriori knowledge of the "world" as well. But in phenomenology,

as Heidegger stresses, quite rightly, the descriptive sense of the apriori as such can be a theme for reflection regardless whether the point of departure is subjectivity or objectivity. The ideation, or intuition, of an "objective" apriori by no means entails an initial turn to the subject as a prior set of conditions for the apprehension of a given object. Such an apriori is instead established along the lines of the distinction, given in ideation itself, of the difference between structural elements of objectivity that are indissoluble from the "this here" of given being and those which are indifferent, thus in that sense "prior," and precisely in such a way that is indifferent to the form of a knowledge of things.[15]

This characteristic of Husserl's approach to the essence of the apriori was already touched on above in the discussion of the theme of ideality in the *Origin*. The experience of ideality is the experience of something that is indifferent to its being-experienced; it has the sense of something that is what it is regardless of the manner in which it is experienced, understood, or talked about. In this sense, Heidegger is correct when he says, "phenomenology has shown that the apriori is not limited to subjectivity, indeed that in the first instance it has primarily nothing at all to do with subjectivity."[16] But when he goes on to say, at the end of the same paragraph, that "the apriori is not only nothing immanent, belonging primarily to the sphere of the subject, it is also nothing transcendent, specifically bound up with reality," he passes over what for Husserl is the key issue: that is, the manner in which we can say that we are "conscious of..." the apriori as such. To be sure, for Husserl, the apriori, or ideality in general, is not a "real" part of the world. Nevertheless, it *is* something "transcendent," which at first need mean nothing more than the fact that it is not an actual constituent of subjective immanence. The transcendence of the ideal has to do with the manner in which it is present to consciousness, thus its givenness as something that can be spoken about in terms of its "sense" or significance. In the case of the apriori, this includes an indifference not only to the flow of conscious life, but to the situatedness of things in the world in general. For the apriori is prior not only to consciousness, but to the world; this is part of its sense, or that indifference of which we are aware in our experience of the apriori.

Yet precisely insofar as it is approachable as a sense, thus as something about which we can speak, and which can be a given for the intuitive act of ideation, the apriori is nevertheless individuated—and precisely in ideation, in the act of grasping the apriori. It is grasped, or experienced "as" that which is neither a part of the subjective nor the objective, but which in its being grasped is nevertheless given a "place," because only in this manner can it "be" for a consciousness at all—or, to use Husserl's language, can it be given in an "evidence" in which we "have" the apriori itself. In this way, Husserl can argue, as he does in §58 of *Formal and Transcendental Logic*, that the evidence of ideal objectivities is *analogous* to real objectivities—analogous, but not identical:

> In just the same fashion, we say, there belongs to the sense of an *irreal* object the possibility of its identification on the basis of its own manners of being itself seized upon and had [*Selbsterfassung und Selbsthabe*]. Actually the effect of this "identification" is like that of an "experience," except that an irreal object is *not individuated in consequence of a temporality belonging to it originally.*[17]

I will leave aside the task of giving a positive account of the temporality that genuinely belongs to ideal objectivities, and stress instead that the apprehension or reflection on logical objects has a complex structure, thanks to which the quasi-individuation of logical objectivity is possible. Logical reflection does not take the simple form of a procedure whereby we first bring a logical object "into view" and then elaborate its structure, characteristics, and properties. "Seeing" here is not a simple apprehension that would present an opportunity to describe what is seen, as if there were, pre-given, an opportunity to appropriate something, to assimilate it as something known. Instead, the kind of seeing at issue already rests on the accomplishment of a particular kind of appropriation; it is already something that takes place in the mode of an assimilation. The ideal, in its being intuited, has already been appropriated or assimilated as something experienced, or at least in "something like" an experience—because the ideal object itself, in its transcendence, is nothing like an object of experience pre-given before its appropriation. There are no raw, unassimilated essences pre-given in experience. And this is something we understand, insofar as we have a sense for the ideal; we understand the necessity for this strange kind of crossing over from the pre-givenness of worldly things to the kind of non-worldly things that essences represent, a passage which governs the possible sense of ideality itself.[18]

Therefore, even if *what is meant* in the ideating intuition of the apriori is nothing subjective, it is still the case that *the arrival at the sense of the apriori* is a subjective accomplishment. The experience of the apriori is founded on the acts that make up a particular subjective orientation with respect to the possibility of experiencing the transcendence characteristic of ideality. In the case of ideal objectivity, transcendence is not given, but it is "assigned," as it were, a "place" by a consciousness that thereby assimilates it in a special mode of subjective experience. But the apriori nevertheless remains indifferent to its own transcendence, experienced from a particular point of view, which also means that it is indifferent to its own objectivity or intersubjective communicability; this is precisely the "sense" or "meaning" of the apriori that, once in place, in turn governs all thinking and discussion.[19] When we speak of the "apriori" we speak of what, in principle, lies before or prior to any thing in the world, any experience, even the ideation of the apriori; this is something that is not compromised when speaking of the apriori as a set of conditions which govern world and experience, for the argument is precisely that the apriori as "prior to" has a unique function or role in the structure of world experience.

The transcendence of the given thing is essentially different: the grasping of the self (*Selbsterfassung*) or the having of the self (*Selbsthabe*) of the thing is the manifestation of a transcendence that is in itself definitive of the sense of the thing. The sense of the thing does not include an indifference to its own transcendence, but is identified with it; the thing is not prior to its own presence within subjective apprehension, but is identical with the unfolding of this transcendence within experience. Still, there is a parallel with the apriori that can be discerned: the thing, as a transcendence, unfolds in consciousness; but this "unfolding" is as much an unfolding of consciousness as it is an unfolding of transcendence in consciousness. So for example when Husserl in *Ideas I* speaks of a parallelism of noesis and noema, this should not simply be thought of in terms of two parallel structures, but rather in terms of a *double movement*. The movement of a thing appearing in perspectives is also the movement of an apprehension in accordance with, or in the mode of perspectival apprehension. The movement is double, but not in the sense that it is a combination of two separate, easily identifiable individual movements; it is rather a difference that opens up within the unity of intentionality itself. The difference between the seeing and the seen, the *intentio* and the *intentum*, can be expressed in terms of the movement of perception that orders itself as an inner doubling of immanence into immanence and transcendence "in" immanence. In this self-ordering of perceptual movement, subjective immanence accomplishes a directedness towards the given, which is precisely the condition for the given to be given as or within its sense, as meaning; consciousness is "sense-giving" in this manner, namely, insofar as in its being directed towards the given it at the same time fuses this directedness with the patterns of sense. Directedness takes many forms, each of which is a "turning to" the given in a particular fashion; and these forms can also build one upon the other, where lower, more basic forms of directedness found higher, more complex, mediated forms of directedness.[20] But at any level, the accomplishment is one of a turning and discovering within immanence the capacity to take up implications within the immanent that have nothing to do with the immanently given alone, but which point towards a givenness that is not its own; in this way, a lived experience unfolds the possibility of the givenness of transcendence. *The apprehension of the apriori is in principle no different*. It, too, is accomplished on the basis of a turning towards, a following through implications found within immanence that are revealed by a shift of direction, of attitude.

Though, to be sure, in the case of ideality the "direction" of intentionality has a complicated, founded structure; it is not simply a turning towards transcendence by discovering within immanence a ready-made implication that only needs to be followed through. It is, rather, the discovery of the possibility of turning towards something that is neither transcendent nor immanent, but prior to both. The argument is, as Heidegger stresses, that "subjectivity" as immanence and "objectivity" as transcendence do not constitute an either/or from the perspective of the problem of meaning. The apriori is a testament to this: for here subjectivity discovers a theme that requires a point of view independent of both immanence and transcendence. Or, better, what the subject has already begun to discover in the capacity to experience a world is radically re-affirmed with the apprehension of the apriori: the difference between "subjectivity" and "objectivity," between the *intentio* and the *intentum*, seeing and seen, is not identical with the difference between an "inside" of immanence and an "outside" of transcendence.

The subjectivity of the world

With this in mind, we can begin to make sense of Husserl's conception of an "apriori of the lifeworld." It is a "subjective" apriori, but that does not necessarily mean that it is the apriori inherent to the subjectivity of the subject, or to its "subjective immanence." The subjectivity relevant to an analysis of the apriori of the lifeworld does not represent a sphere of being circumscribed by the "inside" a subject. Yet the theme of subjectivity is essential to its elaboration, even if it never amounts to the attempt to anchor the problematic of the apriori in the structures of an epistemological subject. It is a subjectivity that should be, I would like to propose, understood as the *subjectivity of the world.*

When Husserl in the *Crisis* writes of a meaning that is "subject-relative," he does not mean the relation of sense to the subject, excluding the given world. He means the present world, as it gives itself to those who experience it. The relativity of subjectivity, the presence of transcendence relative to intentional consciousness, is more an opportunity that the world has taken in order to be manifest than a set of conditions that must be met in order for the world to be given the way that it is. What is here meant by "world" is still a kind of manifest order, but one that is also fluid, for subjectivity is not meant in the sense of something that would fix the world once and for all in a set of conceptual restraints, apriori or not; it is the ever present functioning of relativity itself as the mode in which the world orders itself in experience. To be sure, this order is not yet a fixed structure, in that its forms are not "exact"; the understanding which grasps things within this horizon of relativity does not yet grasp anything like scientific "truth." Yet even when the structures are fixed, when precise measurements are made, when the world is thereby re-appropriated as the sum total of truths "in themselves," the world still remains the "same" unfolding order of the given, where this "unfolding" character represents precisely its "subjectivity."

> As life-world the world has, even prior to science, the "same" structures that the objective sciences presuppose in their substruction of a world which exists "in itself" and is determined through "truths in themselves" (this substruction being taken for granted due to the tradition of centuries); these are the same structures that they presuppose as a priori structures and systematically unfold in a priori sciences, sciences of the logos, the universal methodical norms by which any knowledge of the world existing "in itself, objectively" must be bound. Prescientifically, the world is already a spatio-temporal world; to be sure, in regard to this spatiotemporality there is no question of ideal mathematical points, of "pure" straight lines or planes, no question at all of mathematically infinitesimal continuity or of the "exactness" belonging to the sense of the geometrical a priori. The bodies familiar to us in the life-world are actual bodies, but not bodies in the sense of physics. The same thing is true of causality and of spatiotemporal infinity. [These] categorial features of the lifeworld have the same names but are not concerned, so to speak, with the theoretical idealizations and the hypothetical substructions of the geometrician and the physicist.[21]

It is the same world. But what "the same world" means here is not a common factor shared by the subjective lifeworld and objective science, as if, despite the subjective-relative character of the lifeworld, truth has nevertheless been accidentally discovered even on this level, perhaps in what the Port-Royal logicians used to call a "rash judgment." Rather, it is the "same" world because, in the background of any and all objective formations, all understanding, there remains the horizon of the subjectivity of the world so understood. Or better, not a background, but an irremovable reference to the relative, which is in turn the ultimate horizon in which reference receives its context, thanks to which it becomes significant. The world is thus "subjective" in the sense that it is always relative to an unfolding fabric of articulations in which it, as "world," finds opportunities for self-revelation, appearance, and manifestation. For science, like any formation of understanding, of sense, is "meaningful" only given this backward-reference to the subjective life, or the projects and interests that belong to the world (not only to the subject):

> Just as other projects, practical interests, and their realizations belong to the lifeworld, presuppose it as ground, and enrich it with their activity, so it is with science, too, as a human project and praxis. And this includes, as we have said, everything objectively a priori, with its necessary reference back to a corresponding apriori of the life-world. This reference-back is one of a founding of validity [*Geltungsfundierung*]. A certain idealizing accomplishment is what brings about the higher-level meaning-formation and ontic validity of the mathematical and every other objective apriori on the basis of the life-world apriori.[22]

There are here two questions that must be carefully separated. (1) The first is what the "founding of validity" (*Geltungsfundierung*) means with respect to the apriori. The accomplishment under consideration here is that of the "sciences of the logos," or those sciences which articulate the apriori as a set of idealized structures—thus in the form of the "objects" (*Gebilde*) of the pure logico-mathematical sciences. On this point the passage just quoted seems to be paradoxical: if the accomplishments of higher level meaning formation are an advance over lower level or primitive meaning formation, then in what sense is the validity of what is more advanced (here, more "objective") based upon the more primitive—i.e., where "primitive" does not mean more basic or simple, but inaccurate, imprecise, and relative? Is not this development being presented rather backwards—does not the more primitive find its validity (or not) in the higher order accomplishments, and not vice versa?

The argument of the foundational character of the lifeworld is based upon the thesis that, because of the irremovability of the subjective, any higher order formation carries with it an inherent ambiguity of sense, at least if we look at it from an objectivist point of view (which seeks to remove all the ambiguity of the "merely relative," thus which does not see in it a clue to understanding the character of the whole). The task is to understand the manner in which the higher order structure contains, with respect to the question of its validity, the presence of a subjective moment. One way to think about the problem is to recall a similar ambiguity at work in the concept of telos. A telos is an "end," a "for the sake of which" that, in part, represents a consummation or conclusion, a fulfillment. But all of these definitions also refer the telos backwards, as it were—not back to something wholly other, but rather back to itself. Any movement contains in itself a kind of backward reference that, if it were to cease, would mean that the movement is no longer a movement, no longer "in" the process of being itself. The telos, as an end, or a "for the sake of which," has this sense only from within the course of a movement; this movement in itself, as a realization *in actu* of the "end," is in turn a coherent whole, but not because there is something outside that makes it a unity, but because that which does make it a unity—the telos—is given in the movement itself. The end has its meaning, or presence, *as an end* only in the progression towards it. Or better: the movement is the manner in which the telos is "in effect," "becoming actual," in short: *real*. If so, then to remove the movement from the telos would render it *non-real*, at least as the reality of a telos.

Insofar as the apriori articulated by the logico-mathematical sciences is the accomplishment of an activity, then this apriori, as the telos of its own actualization in scientific thought, is "in effect" only from within the movement itself. It is ultimately (if not *completely*, to again recall the *Origin*) meaningful, or "valid," only with reference to the movement of its actualization. The apriori of the lifeworld is thus not a separate, distinct apriori from that of the sciences, just as the lifeworld is not a separate, distinct world from the world as it is understood in science. Instead, the lifeworld is that to which higher order meaning formation "refers back" just as any accomplished telos "refers back" to the movement that gives it its reality.[23] There is thus an ambiguity, or what I have been calling a *tension*, within the apriori itself, precisely as something "actual" or "individuated" within the movement of a world understanding. To make this tension thematic, or to describe the structure of this movement as a movement, is to make evident not only an historical apriori, but the historical *in* the apriori as well.[24]

This is true for any accomplishment of sense, not just that of the apriori. This leads to the second question, which must be kept separate from the question of founding. Namely, (2) what about the distinctness of the apriori, which has already been described as its "indifference" to its own individuation? We first need to stress that Husserl's concept of the lifeworld is itself a philosophical construct, a theoretical *Gebilde*. It is not, I would like to suggest, a cry to return to concrete life, to reject the hubris of scientific precision and complete knowledge. It is a succinct articulation of two interrelated theses: (a) the full scope of the theoretical, which for Husserl is the bounds of the apriori as such, can include the subjectivity of the world, or the unfolding of transcendence. The claim is that not only the structures of world experience, but also the movement that accomplishes these structures, falls within the domain of the apriori in the most general sense. In this way, a true "logos of the world" is possible, one that would capture not only its objective structure, but the patterns of the lived understanding in which this structure emerges. Husserl's point is that if science, as theoretical life, is to be able to make thematic its own meaning, its own telos, then it must become aware of itself as an accomplishing lifeworld; and his argument is that this is a possibility inherent to the essence of reason itself.

The second, related thesis is (b) that the original givenness of the world—its "pre-givenness" (*Vorgegebenheit*)—is not only the institution of a world "to be understood," or "to be spoken of," but is also the originary institution of the apriori itself. "World" and "apriori" are indissolubly linked within the subjectivity of experience, to the extent to which this subjectivity belongs as much to the world as it does to the subject; though, to be sure, this link is only visible from within the theoretical attitude, and even then only from within a theorizing that has broken through to phenomenology. The indifference of the apriori cuts it off from the world only when we consider both as static objectivities. As pre-objectivities, both find a common set of coordinates in subjective movements which remain, however hidden, even when the line is crossed into fully constituted objectivity. The result is that there is always a fundamental, implicit commonality between apriori and world, a common origin that holds the two together even when they appear to be most separate. Thus the analysis of reason as a movement, as an accomplishing "life," is not merely a perspective on the constitution of thought within the horizon of the lifeworld. For Husserl, it is ultimately the discovery of the problem of the *origin of the world* as such, which is at the same time intimately bound with the problem of the *apriori of the world.*

With these two theses, we have reached the threshold of phenomenology.[25] The problem of the lifeworld is a "way into phenomenology" in that it poses the problem of the ground of knowledge in terms of a subjective apriori that is at the same time the apriori of the world—both of which are the goals of the development of the theme of the subjectivity of the world. The world itself, as the pre-given horizon of activity, is marked by an irremovable subjective character; anything occurring within this horizon has the characteristic of an accomplishment, thus a meaningfulness which in itself, insofar as it is a worldly movement, carries with it its own form a priori.

The concept of the lifeworld thus introduces the theme of subjectivity in a unique fashion; for it is not from the beginning defined in terms of an interiority of a subject, an "inside" posed against the "outside" of the world, but is instead an inwardness inscribed in the figures of the world itself. The world, in a more basic sense than any objectivity that would be defensible as something "fixed" and "true," retains its autonomy *in its own subjectivity*. This can be described in terms of an autonomy of a particular kind of question in the face of the answer it itself has motivated. There are questions that are simply deflated by answers, to the point where one wonders why the question had been posed at all; then there are questions which only become deeper when they are countered with an answer, where any given answer disappears as one more profile, one more *Abschattung* within an infinity of answers that is in turn the only possible correlate of questions of this type. This is what I have been calling the "subjectivity of the world"—a reticence, or reserve, in itself irremovable and never superseded, but only enriched by the meaning, or the understanding it motivates.

This also means that to focus on the subjective, to trace all understanding and sense back to its manner of givenness within the subjective, is not to reduce what is outside to what is given inside an individual being called "subject," "I." Objectivity as some autonomous exteriority is not a reserve against the encroachment of a subject-relative point of view that seeks to dissolve it in a totalizing experience. It is rather the other way around. Subjectivity is a radical exteriority that is never outside, but inside all things; above all, it is inside their objectivity, precisely as that reserve in the face of accomplished sense which never allows understanding to simply stagnate with the "given" pure and simple. That the given is always something subjective for Husserl, always something accomplished within subjectivity, means that "givenness" itself is always a movement—always a "giving of" sense, a *Sinngebung*.

Phenomenology is the study of the many-leveled structure of this movement of sense-givenness, which *in its subjectivity* is the movement of the world. But first this movement must be brought to intuition, brought to view, in its most genuine aspect—that is, precisely in its subjectivity. The theme of the lifeworld represents the first step, but it still leaves us within the circuit of a description of natural life. Subjectivity is presented here in its aspect as a kind of depth of the world; but what this depth accomplishes, the movements that generate its structures and processes of achievement, are not yet before us. To make this happen, we need to engage in reflections on phenomenological methodology.[26]

Before turning to the topics of epoche and reduction, however, I would like to make one more comment. The problem of the lifeworld is not just one way among others to bring us to the doorstep of phenomenological philosophy. It represents a decision about the proper object of philosophy itself, about what needs to be thought, or what calls for thinking. Husserl could perhaps be considered one of the last great philosophers of the Enlightenment, and the *Crisis* his grand defense of reason. But for Husserl, philosophical thinking was not limited to the task of the justification of reason, or at least not a justification that would stop after having found a proper articulation of the bounds of the rational. One could perhaps argue that such an articulation has long since been achieved, perhaps along the lines of the work of Hans Blumenberg.[27] Nor is thinking for Husserl a question of establishing norms for life, as if we could come to life from the outside, or as if the philosopher could ever be a suitable guide for culture and civilization. Rather, what calls for thinking from Husserl's perspective is the given life of reason itself, the fact that life has already formed itself around the pole of the question of its own rationality, its own being-understood in clarity. That which calls for thinking is the fact that our needs are already those of rational beings, including that peculiar need discussed above: the need to be what we are, to be believers who are able to believe.

The phenomenological reduction is a presentation of what Husserl considered to be the methodological framework necessary to meet the task of thinking, understood in this sense. Behind it lies the conviction that it is the essence of reason, as a possibility of life, which is the decisive question of our times.

Notes

[1] In what follows, I will limit myself to following the implications of this tension as it appears in the situation of the crisis of the sciences, and will argue that, already here, the lifeworld as a philosophical theme is possible *before* phenomenology proper. Nevertheless, this does not mean that the full significance of the lifeworld is necessarily apparent. Thus the

thesis of E. Ströker is correct, namely that the lifeworld as *ground* or *foundation* "gar nicht außerhalb einer spezifischen Fragestellung der transzendentalen Phänomenologie vorfindlich werden kann." ("Geschichte und Lebenswelt als Sinnesfundament der Wissenschaften in Husserls Spätwerk," in: *Lebenswelt und Wissenschaft in der Philosophie Edmund Husserl's* (1979) 117.)

[2] On "objectivism" vs. "transcendentalism," cf. Crisis/Hua VI §14.

[3] A similar conception can be found in FTL/Hua XVII: the problem of transcendental logic, oriented around the question of the manifold sense of evidence or truth, is introduced by way of a contrast with the manifestation of evidence articulated in formal logic. And the contrast here has a ground very similar to the one presented in the *Crisis*: there is a resistance on the part of truth itself, to truth as it is lived, to being fully articulated in terms of a *mathesis universalis*, even if such a mathesis can offer itself as a total, systematically complete articulation of the manifestation of evidence as such. Reason itself resists, on some basic level, its own full logical elaboration that would remain purely "objective." Yet it is its putative *completeness* that marks logic as a candidate for such a contrast with the transcendental perspective, which raises an important question: if for technical reasons "objective" formal logic, or formal systems along the lines of Whitehead and Russell's *Principia Mathematica*, can never in fact articulate their own completeness in rigorous terms, does this not disqualify them as candidates for such a contrast? Is the way to the lifeworld through logic barred, precisely due to the fact that the status of logic as an accomplishment, or better a success, has become questionable? For a more detailed consideration of Husserl's logic in light of the history of mathematical logic, see in particular D. Lohmar, *Phänomenologie der Mathematik* (1989); S. Bachelard, *La Logique de Husserl* (1957) 101f; and J. Cavaillès, *Sur la logique et la théorie de la science* (1947).

[4] If so, then Hans Blumenburg's thesis (cf. "Lebenswelt und Technisierung unter Aspekten der Phänomenologie," in: *Filosofia* 14, (1963) 855-884) that the lifeworld is the historically "original" world before the advent of science would be misleading—both the "world" of pre-science and the "world" of modern science are just as "lebensweltlich" as they are "historical." Here one can note how the use of the concept of the lifeworld is similar to that of history: both are powerful instruments for articulating the tensions within a given ("pregiven," "traditional") unity of meaning.

[5] Cf. D1 13a: "Das Sein jedes Realen ist Sein durch Antizipation, Sein aus einer jeweiligen Gegebenheit, die notwendig zugleich Vorgegebenheit ist, und das in infinitum."

[6] Heidegger, on pp. 78-80 of *Einführung in die Metaphysik* (1966), in order to regain a perspective on the hidden ontological dimension of the doxic, stresses that the Greek word *doxa* stems from the verb *dokeô*, "to appear" or "show oneself." Following the study by Sprute (*Der Begriff der Doxa in der platonischen Philosophie* (1962)), I would like to highlight another historical aspect of the word, namely its original association with both *question* and *speech.* So for example in both of its earliest appearances in Homer (*Iliad* X 324 and *Odyssey* XI 344), the context involves some form of a question being addressed in speech. Both play an important role in Plato: see *Republic* 523a and 525 a-b, where Socrates argues that it is the *questionable* in the doxic, experienced in perception, that most induces the soul to thinking, thus ultimately to knowledge (*epistêmê*). And at *Timaeus* 37b-c, the world-soul is described as speaking (to itself) whatever it is that occurs—whether what occurs belongs to the sphere of becoming or to the sphere of the unchanging. The soul's articulation of the former gives rise to *doxai* the latter to *epistêmê*. Philosophical learning, in the form of dialectic, thus in the form of a speech that follows the breakthrough of question-

ing to knowledge, reflects precisely this double self-articulation of the world as such, and in the context of the *Timaeus*, of its *life*.

[7] Cf. Crisis 125, 155-156; Hua VI 127:36-128:10, 158:18-24.

[8] And, following the results of the discussion of the *Origin*, it should also be stressed that such an acquisition is historical. This is compatible with the thesis of T. Tani: "Thus, when Husserl says that the lifeworld precedes science, or that science presupposes the lifeworld, clearly what he means is not that science presupposes a non-historical and primitive world, but rather that every particular theoretical construction of science presupposes its particular historical lifeworld." "Life and the Lifeworld," in: *Husserl-Studies* 3 (1986), p. 60.

[9] Another way to express this is that the openness (*Offenheit*) of the world remains in the wake of its given explication, whether in science or any other human activity. Cf. Hua VI 299:8-17.

[10] Nevertheless, what is perhaps the most fruitful discussion of modality in analytic philosophy—namely, the work of J. Hintikka, S. Kripke, and others—received its impetus from problems unique to the semantics of statements immediately related to perceptions, such as "I see X." Cf. J. Hintikka, "On the Logic of Perception." In: *Models for Modalities* (1969). Also cf. C. Harvey and J. Hintikka, "Modalization and Modalities." In: *Phenomenology and the Formal Sciences* (1991).

[11] This is contra the Brentanian thesis that all modality (all "belief") takes the form of judging activity directed at what is presented (*vorgestellt*). Husserl: "Glaube und Abwandlung des Glaubens ist nichts zu den Intentionen Hinzutretendes. Die Ungehemmtheit und die Gehemmtheit durch parallele und partiell sich deckende Inentionen, das ist nichts neben den Intentionen, nicht ein hizutretendes neues Erlebnis, genannt Glaube, Urteil, sondern eben eine Umstimmung, eine Abwandlung, die das Wesen des Bewußtseins als Bewußtseins ermöglicht, und die in der Tat, wie wir hören werden, jedes Bewußtsein ermöglicht." Hua XI 226:32-39; cf. Hua XI §6 and EJ 90; EU 97:7-11.

[12] This is only meant to be a sketch of Husserl's analysis, which is far more subtle. See Hua XI §§7-8, EJ 89-90, 92-93; EU 96:3-36, 100:14-101:25.

[13] EJ 20; EU 12. One can also point to an even more fundamental distinction that already appears in Ideas I, and which is of central importance for genetic phenomenology: the distinction between originary givenness (*originär Gegebenheit*) and self-givenness (*Selbstgegebenheit*): the former is the givenness of originary experience in its most basic sense (perception), while the latter already involves, even on the pre-predicative level, the functioning of an interest and synthesis. See Hua III,1 81:3-8, 90:2-5, 142:19-25; Hua XI 4:16-17, 6:36-7:5.

[14] This is what Heidegger attempts to articulate in his 1929 inaugural lecture (which Husserl attended), "Was ist Metaphysik?" Dasein is described in this text as the *Hineingehalten*—the holding forth into—within this silence, or a "nothing" (*Nichts*) that is not nothing, but Being in its separateness from beings. To hold oneself forth into nothing, into this silence, is to place oneself in that distance from all that is, within which the question "why?" is itself first possible: "Nur auf dem Grunde der Verwunderung—d.h. der Offenbarkeit der Nichts—entspringt das "Warum?". Nur weil das Warum als solches möglich ist, können wir in bestimmter Weise nach Gründen fragen und begründen." (Was ist Metaphysik? (1965) 24)

[15] The apriori is thus freed from the formalism of knowing and is instead located in the givenness of the known. Cf. M. Scheler, *Der Formalismus in der Ethik* (1966) 68-69: "*Nicht* also an die *Sätze* (oder *gar* an die Urteilsakte, die ihnen entsprechen) ist das Apriori gebunden, etwa als *Form* dieser Sätze und Akte (d.h. an >>Formen des Urteilens<< aus denen Kant seine >>Kategorien<< als >>Funktionsgesetzte des Denkens<< entwickelt);

sondern es gehört durchaus zum *Tatsachen*sphäre, und ein Satz ist nur insofern apriori wahr (resp. falsch), als er in *solchen* >>Tatsachen<< sich erfüllt."

[16] Heidegger, History of the Concept of Time (1992) 74.

[17] FTL 156; Hua XVII 164.

[18] Nevetheless, in an unpublished manuscript from 1917 or 1918, Husserl argues that this crossing over does have a kind of basis in perception, amounting to a mode of implicit presence on the part of ideality in perceptual apprehension. This basis is pre-constituted in such a way that is irremovable from the genetic constitution of perceptual consciousness itself, insofar as the latter contains a structural moment of repetition. Insofar as consciousness is inherently capable of *return*, where return also leads to associative syntheses, the very presence of consciousness is the implicit formation of ideality, even before this ideality is made into an object. Ideality is not thereby "experienced," rather the experience of consciousness marks out a "singularity" of the ideal, as a preparation for its individuation. See D5 4a/b: "Jedes gemeinsame Wesen ist bei den synthetisch verbundenen Akten bzw. den intentionalen Gegenständen 'wiederholt' da; das Deckungsbewußtsein ist ein Gleichheitsbewußtsein eventuell ein Bewußtsein vollkommener Gleichheit. [...] Das ist nicht Wesen im Sinne von Eidos, vielmehr die Vereinzelung des Eidos, also das Wesensmoment, und nur für das vollständige Wesen als Moment haben wir keinen Namen. Dabei kommt es auf keinerlei Attentionalität und überhaupt auf keinerlei Vollzug von aktiven Spontaneitäten an. Das Eidos ist zwar 'da,' aber es ist für das Bewußtsein nur da durch seine originale Konstitution, für die hier alles vorbereitet, aber noch nicht alles geleistet ist."

[19] Ideality thus "steps forth" (*auftritt*) in a thinking that does not thereby signal the individuation of the ideal. Cf. EJ 263-264/EU 316:36-317:3. (Churchill and Ameriks translate "Auftreten" with "appearance," which should be taken in the colloquial sense in which someone "makes an appearance.")

[20] (a) On the notion of a pre-egological directedness that belongs to the passive sphere, see Hua XI 77:27-37: "Ich sagte: Retentionen, wie sie in ihrer Ursprünglichkeit auftreten, haben keinen intentionalen Charakter. Das schließt nicht aus, daß sie ihn hinterher annehmen können unter gewissen Umständen und in ihrer Weise. Wir dürfen uns also nicht dadurch verwirren lassen, daß uns gelegentlich gerichtete Retentionen entgegentreten, wie in jedem Fall, wo das Ich auf ein Retentionales seinen Ich-Blick richtet. Denn allgemein gilt, daß ein Vorstelliges, auf das das Ich seinen Blick richtet, ein Wahrgenommenes, Erinnertes, auch ein Retentionales, schon in sich selbst intentionales <sein>, also schon in seinem passiven Gehalt Richtung auf sein Gegenständliches haben muß." Cf. C6 6a: "Wir haben also 'Bewußtsein von' in verschiedenem Sinn und in verschiedener Fundierungs-stufe: (1) Das ursprünglichst zeitigende Bewußtsein, in dem durch die Urimpressionen (Urpräsentationen, Urretentionen und Ur-protentionen) sich 'immanente Daten' (die Empfindungsdaten, mit ihren Gefühlsmomenten, auch die Triebmomente, alles in der Weise des innerst Zeitlichen) konstituieren, aber auch die Erscheinungen-von, die Ichakte, alles, was überhaupt zeitlich-innerzeitlich eins ist. (2) Akte. Das eigentliche Bewußthaben, darauf Gerichtetsein des Ich. Das Bewußtsein sub 1) ist uneigentliches Bewußtsein, ist keine 'Intention.'" (b) The directedness of the ego, itself in part founded on passive directedness, is essential for the formation of "objectivity" in the genuine ("intentional") sense, whether it is perceptual objectivity or the objectivity of objects of the understanding (*Verstandesgegenstände*). See EJ/EU §§16-19, 47-49; Hua XXXI §§1-6.

[21] Crisis 139-140; Hua VI 142:22-143:3.

[22] Crisis 140; Hua VI 143:9-19.

[23] This reference takes place within knowledge itself, along the lines of a tension between implicit and explicit. M. Dufrenne: "An implicit knowledge appears as present in me without needing to be formulated; this is why it is important to distinguish the apriori from the pure knowledge which makes it explicit and which is not always attainable (or at least is never complete): whereas a primordial certitude is always present. In order to be actualized, this certitude does not have to equal truth; it suffices that it found truth." *The Notion of the Apriori* (1966) 122-123.

[24] The interpretation I am pursuing here is not meant to contradict more standard readings of the *Crisis*, represented by P. Janssen (*Geschichte und Lebenswelt* (1970) §26), which, taking as its point of departure the passage at Hua VI 142:16-143:30, where Husserl argues that the lifeworld has a "universal type-structure," interprets the apriori as an invariance of types or forms of doxic belief. Cf. also Hua IX 220ff. I am instead proposing a shift of emphasis away from typology and towards the theme of the movement of world understanding—the idea being that the latter allows one to better understand how the lifeworld, precisely as a thematization of the historical essence of the apriori, offers a "way into phenomenology."

[25] And in doing so perhaps evaded Gadamer's ontological "either/or": "*Entweder* gibt die >reine Wahrnehmung< die Orientierungspunkte ab, auf die am Ende die Naturwissenschaften aufbauen (und die sie in Wahrheit durch Experiment und Messung allererst herstellen). Das ist die Ontologie der Vorhandenheit. Dann hat die Rede von Fundierung Sinn. Oder man erkennt darin eine unphänomenologische Konstruktion, die einen derivierten Modus vom In-der-Welt-sein priviligiert. [...] sehe darin zwei verschiedene Ontologien, die der griechischen Metaphysik und die durch deren kritische Hinterfragung geöffnete Stellung der Seinsfrage in Horizont der Zeit." Hans-Georg Gadamer, *Gesammelte Werke* 3 (1987) 148n. First, because it is not the world of "pure perception" as such, but the historical world of its own emergence with which science is in tension; thus the re-orientation represented by the lifeworld is not necessarily in direct conflict with either experiment or measurement—but is precisely the thematization of the world-opening capacity of both. Second, because this tension is still only a nascent form of a "fundamental questioning" that has not yet taken—nor may ever take—the form of a *Seins*frage. It is still a *Welt*frage, moving within the horizon of the questionability of the world.

[26] Such methodological reflections move beyond what Husserl, in Crisis/Hua VI §51, refers to as the "ontology of the lifeworld." Thus H. Lübbes' argument ("Husserl und die europäische Krisis," in: *Kant Studien* 49/1 (1957/8) 334-335) that Husserl neglects this ontology in favor of transcendentalism is misleading—insofar as one accepts that the task of phenomenology is above all the clarification of subjectivity itself. To understand an ontology, even of the lifeworld, as such a clarification would be to accept the analysis of a product of subjectivity as a substitute for the analysis of subjectivity itself.

[27] Blumenberg, *Die Legitimität der Neuzeit* (1976).

Chapter Six

The Phenomenological Reduction (*Crisis* §§ 35-55)

Epoche, conversion, and still-life

If the phenomenological epoche is a device that will bring what in the last chapter was called the "subjectivity of the world" into view as an object of inquiry, it does this only insofar as it is an exercise of possibilities inherent to that same subjectivity.[1] That is, the epoche is to make visible, as a theme for reflection, the world itself in its unique openness to its own meaning, and this is possible only because of this openness itself, as a function of an accomplishing subjectivity. Guided by the theme of the lifeworld, what we pursue initially in the epoche is a very different matter from the pursuit of understanding the world as such, establishing "what there is" in a direct comprehension. The epoche is, so to speak, designed to circle around, even to avoid, the explicit accomplishments and possibilities of such an understanding. It sets up a reflection on the world that no longer seeks to bring the world into the clarity of something understood, but remains in a kind of twilight of comprehension, where the world is not yet understood and the understanding is not yet identified with a subject, insisting that in this twilight there is still something to be said. In the *Crisis*, Husserl's argument is that such a short circuit, or "bracketing" of the world in its guise of something comprehended, is an inherent possibility of world experience; it is in fact the possibility of subjectivity itself, initially interpreted as the openness of the lifeworld to its own being-comprehended.

It is important to stress from the beginning that this reflection remains, at least at first, within the world, though now within the further limitation of the subjectivity of the world. Reflection remains in the world because of the simple fact that the epoche, like any act of thinking or understanding whatsoever, occurs in the form of an attitude, or a directedness towards such and such a theme. This attitude, like any other, finds its place within the horizon of the world. It finds a unique position within the whole in relation to which, for the sake of the activity at hand, other activities count as "excluded." To be excluded, however, is not to be rejected or negated, but pushed into the background to make room for the given concern. Once an activity, any activity, is given such a place, or more precisely has been accorded its own "time"—its own "time for...."—it becomes a readily available part of the world. It becomes a habitus that can be adopted, as well as something that demands our attention as a set of interests and concerns already "in place," which also means "opened up." And such demands fall into place, or are in effect, always with an accompanying exercise of a distance between the activity at hand and the demands and interests of other activities. Thus "epoche," as a name for the exercise of such distancing in general, is not unique to philosophy, but is characteristic of any activity, anything that we do; it is the mode in which an activity takes up residence in the world. And same thing is true of the philosophical epoche that is true in any mundane epoche: the distance we exercise is a distance within the openness that is the world.

In the case of phenomenology, there is a particular concern with creating such a distance, or finding an opening with respect to science:

> Within this epoche, however, neither the sciences nor the scientists have disappeared for us who practice the epoche. They continue to be what they were before, in any case: facts in the unified context of the pregiven life-world; except that, because of the epoche, we do not function as sharing these interests, as coworkers, etc. We establish in ourselves just one particular habitual direction of interest, with a certain vocational attitude to which there belongs a particular *Berufszeit*. We find the same thing here as elsewhere: when we actualize one of our habitual interests and are thus involved in our vocational activity (in the accomplishment of our work), we assume a posture of the epoche toward our other life-interests, even though these still exist and are still ours. Everything has "its proper time", and in shifting [activities] we say something like: "Now it is time to go to the meeting, to the election," and the like.[2]

"Epoche" is thus the manner in which an activity finds its place within the space of the world—that is, within a horizon of "given time," in which the time given affords the possibility of each activity having something of its own, without which each would be impossible. This point is important for Husserl, as he himself stresses, and for two basic reasons. (1)

The first is implicit, but important given the interpretation of the lifeworld presented in the last chapter. If the subjectivity that phenomenology will make its theme is, at least initially, in the form of the subjectivity of the world, then the very activity of the epoche must be understood in terms of the possibility of such a reflection finding a place in the world for the appearance of the subjectivity of the world. Far from being a radical new departure for Husserl, this is in fact a deepening of the presentation of the epoche in *Ideas I*, where the possibility of philosophy is introduced by way of a reflection on the ego as "man in the world." The second reason (2) is explicit:

> There are good reasons for my stressing so sharply the vocational character of even the "phenomenologist's" attitude. [...] most important, the suspension of [the] accomplishment [of the epoche] in no way changes the interest which continues and remains valid within personal subjectivity—i.e., its habitual directedness toward goals which persist as its validities—and it is for this very reason that it can be actualized again and again, at different times, in this identical sense.[3]

This is a theme we have already met—the availability of sense as a persisting validity, as a part of the meaningful context of the "surrounding world" (*Umwelt*). In the discussion above of the phenomenon of tradition, we have also seen that this persistence is not simply the preservation of sense, the establishment of sense that can be passed down through the generations as identically the "same" meaning. More importantly, it is an establishment of sense which orders the world, giving it its particular style and character, effecting both the manner in which questions are posed and answers pursued. A validity that persists, which carries weight with us, in turn reflects the manner in which the sense of the whole is "understood." And if phenomenology, as philosophy, is not only to be something rationally justified, but a validity in this other sense, thus a valuation which gives weight to a certain possible shape in which this sense of the whole can take for life, then the issue here is more than simply the possibility of a repetition of validities already accepted or in place. The question has to do with the possibility of the lived value of philosophy, as an experience that has a place in the functioning of the spiritual whole of life, or how the whole is reflected in its individual components.

It is thus in part because of the importance for philosophy of the question of its own possibility that the epoche comes first in the methodological exhibition of phenomenology proper. In the case of an activity defined immediately by the given concern that gives it direction, where there is no question that the concern is legitimate and not illusory, such an emphasis on "making room" would have no place. Epoche would simply be one among many characteristics of the activity itself. But for philosophical activity, epoche is a specific step that must be taken; one must probe for a place, thus for a concrete possibility; and with that comes a heightened awareness of what it is that the epoche, in any activity, actually accomplishes. This is not that different from the reflection on the methodological use of doubt in Descartes, which Husserl further refines in *Ideas I*. In neither case, Descartes or Husserl, is it a question of actually being in doubt; the point is instead to reflect on potentially interesting possibilities that lie dormant in the positioning of natural doubt, possibilities of which we are only dimly aware when actually under the sway, or in the grip of doubt itself. In *Ideas I*, Husserl presents the reader with an idea of the epoche as the successful disentanglement from the belief that binds us to a thesis—an epoche more subtle than Cartesian doubt which, while no longer natural or even "actual" doubt, remains a kind of temporary self-denial. For Husserl in *Ideas I*, epoche is not denial, but disengagement, holding the thesis at bay, thus making it available for a particular mode of reflection that would not have been possible if the disengagement had not been effected. We thereby, and in a systematic manner, uncover a possibility of reflection that had necessarily remained dormant before the exercise of the epoche.[4]

In the *Crisis* the presentation of the epoche is even more subtle, and more powerful. Here the key is no longer an advantageous aspect of doubt that allows us to free ourselves from a belief in the form of a thesis. To be sure, the epoche is still understood as a kind of disengagement, but it is no longer premised so much on an arbitrary freedom to pull out of the performance of any and all acts of understanding, all "theses" of the standpoint of experience that is "natural" to life. It is now premised on the gradual discovery of tensions within natural life itself. Or better: tensions within the lifeworld, which includes the tension between science and nature, or the "naturalness" of the world of life and the "unnaturalness" of the infinity of modern science. Bringing out these tensions, making them a theme, is already the beginning of the epoche, already a kind of "bracketing." It begins to lift reflection out of naiveté by following the fault line of a fundamental tension between science, as a traditional set of validities that have weight for life, and the question of the subjectivity of life, the horizon in which all validity is set. Nevertheless, a fully exercised epoche is necessary, because this is a tension which must be brought out, it does not lie on the surface as a conflict with which we are "naturally" concerned. On the contrary, it is a tension hidden within the naiveté of objective science itself—a naiveté that, like all naiveté, as was stressed above, is productive, positive, determinative of a particular style of experience and world. Thus it is a tension which, even if hidden, is in place, plays a role, and in turn harbors the possibility of a kind of reflection with respect to the problem of the lifeworld: the dissolution of the naiveté of the objectivist attitude in favor of bringing to the fore the functioning openness of subjectivity that saturates that very same attitude.

The subjectivity revealed by the epoche of natural science—which is nothing but the discovery of subjectivity within the naiveté of science as a human accomplishment—takes a certain form: that of the world-as-lived, grounded in the relativity that is the world-opening function of subjectivity. This "subjective" world is concrete, natural existence. Yet it is not a sphere of clarified presence that had been forgotten by a narrow-minded objectivism. The epoche of science goes hand in hand with a dissolution of a certain kind of naiveté, what could be called the naiveté of the objective; but that does not put any distance between us and a deeper, more basic naiveté: that of the subjective. Naiveté does not mean obliviousness alone, but also a given acceptance that no longer needs to be put into effect. As given, it is already in effect and as such taken for granted. Science is "naive" in this sense—the particular understanding of the world that it represents, "objectivism," is "in effect," taken for granted, where we feel that in science we have something, a grasp of what the world "is." But, Husserl argues, deeper than this grasp of things, and forming its ultimate basis, is a "having" of another sort, which is subjective. On the level of the lifeworld one "has" something even before it has been understood, even before one is able to provide any "objective" explanation. That there is something before understanding in any sense, a context within which knowledge is formed, a having of the world which is more original, a field within which the possibilities of knowledge are set—this is the deeper sense of the subjective, the elaboration of which is the task of phenomenology.

But this deeper sense of the subjective does not directly result from an analysis of the lifeworld. The lifeworld is the world of original evidences, the living presence of things; its concrete, apriori structures are the teleological depth of the objective apriori, thus the ground for all understanding, all grasp of objectivity in the widest sense. Yet this world "is" precisely in the form of the "pre-given," which means that its givenness is characterized by a certain naiveté—again in the sense of an accomplishment, that availability of sense in the form of its having been given. The pre-given horizon of the world, itself a subjective accomplishment, sets into place a given in naïve acceptance; it represents no claim to objective validity, only to the self-validity of subjectivity. It is, in short, the horizon of the self-accomplishment of subjectivity itself, or in other words the horizon in which *subjectivity* has taken *itself* for granted.

All of the evidences of the lifeworld, the whole of its apriori, must in turn be set within the wake of this accomplishment—which is precisely why one can speak of the "subjectivity of the world" at all. To be sure, this raises some questions. Is there not perhaps another hidden tension to be found at this level, analogous to the tension between the objectivity of science and the subjectivity of the lifeworld? Must there not also be a tension between a transcendental subject and the world constituting horizon of subjectivity as such? And if so, then is there not also the possibility of another epoche, or the place for another mode of reflection, that would have as its theme another sense of "subjectivity" different from the subjectivity of the world? Does not the very theme of tension raise the specter of the subjectivity of the subject, understood in such a way that would cut the ultimate, founding sense of subjectivity off from the world altogether?

Yet the way that the theme of tension is developed in the *Crisis* blocks this pitfall. The tensions that are relevant are those which are found only within the experience of the world, thus which are continuously *sustained* by the movement of world experience. A metaphysical tension between a radical interiority of subjectivity that sets itself off against an alien exteriority of the world cannot be understood as a tension that plays itself out within the world, that is, within the subjective accomplishment of the world as horizon of meaning. The fundamental tension must instead be between subjectivity and its *own* world, or to be more precise, between subjectivity and its own self-accomplishment as world-constituting life—thus in the end a kind of tension within a subjectivity that is always already a world. The insight behind Husserl's conception of the lifeworld is that subjectivity is irremovable from the sense of the world, which means that it is meaningless to speak of the subjectivity of a subject as a separate sphere of being from that of the being of the world that would be located in the beyond of some incomprehensible "outside." Still, this leaves us with a problem. How are we to understand the sense of the obvious difference between subject and world, which is definitive even of the intentional relation itself? The *intentum*, the given articulated in its "noema," is to be sure nothing "outside," free from and alien to the subjective, but it is nevertheless defined by a difference with the subjective—not only with respect to its object-core but also its entire structure. Intentionality as such is really only a title for a problem: to understand how this difference between subjectivity and objectivity, the self and the other, can be drawn within a deeper sense of self, a deeper sense of the subjective. How is it that the origin of the world lies within the accomplishment, from within subjectivity itself, of a difference that begins a kind of immanent movement of exteriority?

Furthermore, if such a tension is to be brought out, made into a theme, its articulation given a place through an exercised epoche, then it must take on a different form than the simple suspension of validity. The epoche of science is an example of the latter type of suspension. It was possible because of the tension between the validity of the scientific and that of the pre-scientific. The axis of this tension cut along the lines of a validity of the world as subjective accomplishment along with a countermovement within this accomplishment that seeks to establish the validity of something that in the end is impossible, at least in a complete form: a grasp of the world that would exclude everything relative to its subjective sense. But the tension between subjectivity and world in this more fundamental sense is not the tension between two validities. The world in its pregivenness, thus sustained by the naiveté that characterizes it, will remain the ground of all validity. An epoche here, then, does not simply make room for one validity by pushing another into the background (where the presence of validities are bound to activities that either presuppose them, or pursue them). Instead, epoche is here the progressive awareness of a tension between subjectivity, as a theme for itself, and the ground of all validity as such, the "world" understood as self-unfolding subjectivity. As a consequence, there is no question here of abstaining, step by step, from one validity to another:

> The abstention from performing individual validities (similar to the way this occurs in a critical attitude, caused by theoretical or practical demands) only creates for each instance a new mode of validity on the natural ground of the world; and the situation is not improved if we wish to exercise, through an anticipatory, universal resolve, the abstention from the performance, one by one, of all validities, even to infinity, i.e., in respect to all of one's own or alien validities which from now on could ever suggest themselves.[5]

The simple abstention, even into infinity, from the performance or affirmation of individual validities does not break the circuit of what Husserl, in the same section (§40), calls the "single indivisible, interrelated complex of life"—that is, of natural life, the vital horizon of all validities. To break such a circuit would entail, as Husserl claims in the previous section (§39), a "total change of the natural attitude," a re-orientation whereby one ceases to accept the horizon of the world as the pre-given horizon of reflection.

"Natural attitude" is a concept from *Ideas I*. In the *Crisis*, however, it is no longer a question of suspending a thesis characteristic of natural life, as if the standpoint of the world were analogous to an individual belief. The standpoint of natural life is not that of a thesis, but represents the living horizon in which beliefs are formed in a complex set of relationships, a texture of validities that are in a state of continual (and relative) becoming. The whole of this life is a texture of beliefs; but it itself is not based upon a single belief, or validity, but rather upon an acceptance that has always already placed the subject in a position from which it can believe, thus participate in the sense of the world as a synthetic movement of validities within the pregiven horizon of this same acceptance. This "acceptance" of the event of world and validity gives experience, conscious life in general, its character of "straightforward" involvement—the unquestioned acceptance of bearing the importance of beliefs, of validities, of the task of understanding the world. *This* is what is to be bracketed: not a thesis, but a straightforward movement, the gesture of what it is to engage a world in general. And in doing so, in holding back this gesture in reflection, Husserl seeks to gain a vantage point not *within* the horizon of belief and validity, but so to speak *above* it:

> Instead of [the] universal abstention in individual steps, a completely different sort of universal epoche is possible, namely, one which puts out of action, with one blow, the total performance running through the whole of natural world-life and through the whole network (whether concealed or open) of validities—precisely that total performance which, as the coherent "natural attitude," makes up "simple" "straightforward" ongoing life. Through the abstention which inhibits this whole hitherto unbroken way of life a complete transformation of all of life is attained, a thoroughly new way of life. An attitude is arrived which is *above* the pregivenness of the validity of the world, *above* the infinite complex whereby, in concealment, the world's validities are always founded on other validities, *above* the whole manifold but synthetically unified flow in which the world has and forever attains anew its content of meaning and its ontic validity. In other words, we thus have an attitude *above* the universal conscious life (both individual-

> subjective and intersubjective) through which the world is "there" for those naively absorbed in ongoing life, as unquestionably present, as the universe of what is there, as the field of all acquired and newly established life-interests.[6]

Husserl then goes on to remind us of the description of the epoche in comparison with other "vocational attitudes." This "being above" the whole of conscious life, of the validities of the world, of the pregiven horizon of validity (and belief), of the interests of life itself, is assured a place "in" the lifeworld. But what is such an assertion supposed to mean? If life is the straightforward engagement of interests, of the fabric of validities that unfold a world, then what "place" in this life, or in this world, could such a point of view have? I would argue that this claim has meaning only if we do not take this expression of "being above" life to mean "outside" life, as a perspective on the world that would be independent in the sense of being separate. So then how is it to be understood? What point could such a point of view have?

There is an important clue towards the end of §35, just after the description of the epoche as a mode of *Berufszeit*, where Husserl compares this new attitude to a religious conversion. Like the epoche, a conversion is a shift that transforms the meaning of the whole; in this sense it is not one "vocation" among others, but something that redefines the sense of the whole of life within which all activities are pursued.[7] This is one sense in which the phenomenological attitude is "above life." Nothing, at least with respect to its sense, its value, will be in principle indifferent to what philosophy reveals, insofar as philosophy succeeds in formulating the sense, the meaning, of the ground of all value, of the "world" as a whole. But there is more to this than that. Conversion need not simply mean the introduction of a new meaning, the acceptance of a new set of ideas—to collapse the distinction between conversion and doctrine, or even conversion and faith, would neglect the complexity of the human drama that unfolds between these poles. In Augustine's *Confessions*, faith—the gift of faith—is an ever-present certainty, one that we cannot, under our own power, decide in favor or against. If faith is present, it is simply in place; but that does not mean that this place renders it unproblematic, that we thereby understand how it is that this faith fits into our life. Likewise, when Husserl talks of the world itself as a horizon of certainty, that does not mean that, simply by having the status of certainty, there is nothing left here to be said, as if there were no cause for concern in the face of those ambiguities and questions that, paradoxically, characterize this very "certainty." Something else has to occur for this certainty to be realized as something fully meaningful, as opposed to a source of questioning and despair, or a kind of uncertainty of certainty. Above all, it is not as if a place needs to be found in us, "in our life," for this certainty; it is already there, present, even definitive of what we are. Rather, the situation of faith is that *we* need to find a place, not for it in us, but for us in it. And that means, both for Augustine and Husserl, that we need to find for a place for ourselves in ourselves. *It is in ourselves that we are not at home, out of place.*

The process of finding a place for ourselves in ourselves is, in Augustine, called "conversion." And in one of the most poignant passages of the *Confessions*, Augustine describes one of the initial phases of the conversion process as the experience of being turned around to face oneself—God turns Augustine's gaze to fall back upon his own face, as if there were a space just in front of his life where he can be led around a circuit in order to see, to face what it is that his life is, or has become. This seeing what it is that I am—which for Augustine is first and foremost the recognition of one's despair and wretchedness from a newly found "point of view"—also suggests a new possibility, both for the meaning of the unease, of the despair of life, and the meaning of faith. Augustine describes himself as always having lived in the forward movement of a life defined by both wretchedness and faith, a forward living of a question, or of a life that takes the form of a search, but also of a growing understanding. But the decisive moment is not when he finally matures into a knowledgeable servant of wisdom, but rather when God enables him, as it were, to *live back into* his life, which has been made newly accessible by the circuit he has followed in being turned around in this space just before—just above—the forward movement of his existence. A new chance for living through the meaning of this life, thus for finding a place within the self, is thereby possible, or opened.

A similar structure is found in Husserl's explication of the epoche. Natural life, life in the world, is a forward-moving life in a horizon of validity already in place, the horizon of objects "towards which" we are directed in an interested fashion. This is also the directedness within which validity is established, not simply found or carried over to the present from the past; and with that comes an awareness of an ever-developing comprehension of things in the movement of experience. In natural life we are thus already aware of the world as a horizon of movement, we are aware of both the openness of the world to the formation of new structures of meaning, as well as the movement of comprehension that is our understanding. Natural life is "naive" to the extent that its awareness of itself is fully caught up in its own forward movement—it is committed to the trajectory upon which it has embarked, a commitment that is part of its very motion, like the commitment of a throw, or a speaking.

In the epoche, a moment of this forward moving life ceases, and turns to look—not back at the "world" alone, but at its own movement as the movement of self and world, which is as such a movement of mutual obliviousness, precisely in order to be what it is. I look back at that self which I am, for whom the world is the kind of certainty that it is, because of this intimate relation between the meaning of the world and the being, the "life," of which my self is the accomplishment:

> I turn to look at myself in my certainty of myself as one who is a human among humans in the world, and I want to question this world-certainty, this being and life of mine, I want to question wherein and in what validity levels the world becomes valid for me in all its content, its entire ontological structure.[8]

Husserl's strategy is subtle, and more importantly, non-Cartesian. In the epoche, I do not turn away from the certainty of the world to the certainty of the "I am," but instead turn back upon a certainty that I effect in every moment of my life—in a direct, straightforward manner in the very movement of my existence in the world. "Certainty" here expresses the primordial self-engagement of life. And it is in the wake of this being-directed in such an engagement that something like a horizon can be lived at all, that the world can exist as the universal synthesis of all validity, past and future; and if "being-directed" is a kind of "having," then the universal synthesis "world" is indissoluble from the scope of this having:

> I am saying: "the world is my universal validity," by which is to be understood: a universum of all validities is always already present as a unity-validity in which I am in a positive sense, throughout a course that is both partially happening and sedimented, but also in the course of a new establishing [*Neustiftung*]. And this universal validity-synthesis is the synthesis of aiming intuitions in the movement of successfully attaining their ends, habitualizing, or missing their mark; and so we have a movement of acts and their sedimentation in a unity with their sedimented legacies, where we have the same acts in [a] new mode of preservation [...] This validity-whole and the whole as legacy with the predelineation of the future, and the pre-validities that go into such a predelineation, always

> already has an ontological form—precisely that of *the world.* It is the world that is my validity, and in which I am now; I am the one who is as naïve "agent" of this universal validity, and I am continuously the same I, existing in the universal purposiveness [*Willentlichkeit*], maintaining it by moving forward, in an ontological sense—yet moving forward.[9]

This *Vollzug* or carrying out of the validity of the world, a metaphor that describes the manner in which life has the world in which it moves, has a complex structure. Its elaboration would require a more detailed presentation of Husserl's genetic phenomenology. For my purposes here, I would only emphasize that this accomplishment is described by Husserl as *forward moving*, which is meant as a more precise description of the natural quality of life, of its "naiveté." This forward-movement of subjective world-having, world-life, is the necessary form life must take in order for it to be life on the basis of the world, or in the world it possesses as a basis (*Boden*). The pre-givenness (*Vorgegebenheit*) of the world and the world as ground (*Boden*) are intimately connected within this forward movement [*Fortführung*] of life; in the engagement of things in experience, in being directed to things through an interest, the world takes shape as the pre-given horizon and ground for the movement of interest, even intentionality itself.[10] The two are united within the same movement, for they are two aspects of the same possibility—the possibility for something like a "world," which is in turn identical with the possibility for something like a "life." But this double possibility can be realized only in this directedness, only in an intentionality as that which moves forward. To work against this forward movement is, in an important sense, to work against life, or at least what life is naturally: the tendency to be manifest as a world-being, a world-life.

Which is exactly what the epoche does—it is an activity that, turning towards the most fundamental possibility of life itself as the double movement of self-accomplishment and world accomplishment, *seeks to hold this accomplishment still*, i.e., seeks to arrest its forward motion in the space of a "turning back." And this, Husserl tells us, is ultimately that on which philosophical reflection, understood as *Besinnung*, is based:

> I inhibit this constant universal "accomplishment" in the epoche, which enables me to pursue a universal reflection [*Besinnung*] on my being as an I, as that being which constantly has, had, and will have, insofar as it enacts and

> transforms it, this entire purposiveness, and in so doing establishes this style of the total legacy. In place of being in accomplishing, in forward moving activity, continuously intending, continuously being directed at being that remains in validity from out of what had been earlier intended and realized, and which is woven forward into the being of the world, I hold myself in a certain sense still.[11]

Reflection holds still that which otherwise—"naturally"—flows, and which is therefore invisible. The world of things, in its flowing forward as the experience of life, is visible as a world of things with which I am concerned—it is not visible as an accomplishment of subjectivity. That "flow" of the subjective, the naiveté of natural life, is invisible, a background accomplishment apparent to natural life at most in the guise of a horizon already in place, a world with which we are always already concerned. But in the epoche we hold this flow still, in the sense that we turn around to face it—to face ourselves, or our being as the subject of the productive movement of our life—in order precisely to make this flow itself, as the having of a world, into a question ("nach diesem Haben, nach dem Wie des Habens dieser Habe zu fragen").[12]

As in Augustine, in the epoche the reflecting subject faces itself—it does not cease to be itself, on the contrary. It is on the way to becoming more of itself than it had ever imagined. But this new possibility of self that is discovered is explicitly "unnatural," in that it has turned against the flow of selfhood (*Ichleben*; *Ichsein*). And if *Besinnung* is here a *Selbstbesinnung*, that does not mean that we have discovered the problem of subjectivity as something independent of the world; the central issue is still the pre-givenness of the world, the accomplishing of meaning latent in the naiveté of natural life itself. *Selbst-besinnung*, as a possibility of bringing subjective being into view, is understood by Husserl as the bringing of worldly being into view, that is, the manner in which we have the world in the most universal of senses.

More, as was emphasized above in the discussion of the meaning of *Besinnung*, the point is to *shape* this world-having, to integrate it into life as a theme in itself, thereby taking natural life into a new dimension of itself—re-integrating it back into itself as a whole, a unity of understanding that no longer rests on the direct accomplishment of natural life, on the ground of its own being. This is a deeper sense in which the epoche is a "turning" comparable to a conversion. In the epoche, as an archetypal philosophical gesture, I turn to face myself, my life, thereby gaining the opportunity not simply to be what I am, but to be what I am as the one who is responsible for what I see.

Reduction as phenomenalization

If the *epoche* is the establishment of the place, in reflection, for the appearance of the world-having accomplishments of subjectivity, then the *reduction* is the means whereby we orient ourselves in order to be able see what there is to see in such a place.

To find a place for the theme of subjectivity, in the sense of epoche described above, requires the de-circuiting of the world as pre-given ground. Thanks to the epoche, the pre-given ground of the lifeworld itself is no longer invisible; its functional invisibility is no longer sanctioned by the naïve certainty of the givenness of the world. It now becomes potentially visible in a unique sense once I have, in effecting the epoche, positioned myself "above" this naïve certainty by holding still the forward movement of the accomplishing activity of subjectivity. The full making visible of what was invisible before the epoche is what Husserl refers to as "reduction," which I will describe here as "phenomenalization." It is a kind of articulation of what, after the epoche, one has been put in a position to see.

Visibility in the specific sense of the phenomenality secured by the reduction is not simply a correlate of intuition. The phenomenality of the pre-given is premised on the holding-in-place of something that, in its "natural" rhythm, is never held in place, or bears the stamp of the thematic in the broadest sense. Thus "reduced" means, to a certain extent, "captured," in the sense that a movement can be captured by the sweep of a line in a drawing, or a figure by the juxtaposition of light and dark, as in a self-portrait by Rembrandt.[13] When, standing over natural naiveté, and holding its being-accomplished in place, the world has become "in a quite peculiar sense, a *phenomenon*,"[14] it does not simply appear in intuitive clarity. If it did, then we would not have to speak of a reduction, but could simply lay claim to a new region of subjectivity as if it were another aspect of natural world-life, albeit available only given the peculiar perspective of the epoche. This is the first characteristic of reduction: it aims at a phenomenality, thus a visibility, which is of a different order than the "natural" visibility of the world.

In light of this, the question of the accomplishment of the having of the world is not a question that can be posed within experience that has been formed, and given its sense, in the wake of the event of the world in its pre-givenness. The very acceptance of the world as pregiven blocks such a question. It does this not because it forbids the question to be asked, but because it "gives" the world in such a way that this question can have no sense, no motivated interest to support its significance as a question. The world as pre-given is precisely a world that does not need *to be* accomplished; it is always already "had" by us, and does not appear to us as an accomplishment, but as a horizon *for* accomplishment. Thus when Husserl argues that the epoche, which de-circuits this pre-givenness, allows the question of the originary having of this horizon of the world as an accomplishment to be posed, he is not claiming to have discovered an important, thus meaningful question that has been neglected. He is making this question meaningful *for the first time*. And what thereby becomes questionable is not the "world" in a direct sense, but the world in a sense for which we have always had a blind spot, precisely due to the forward movement of our lives. For the originary having of the world to be questionable for us, we must first learn to see where we are "naturally" blind. In many ways this amounts to looking just past the fold of natural vision into the horizon within which it unfolds; and in doing so, we look past the visibility of the whole, in order to capture in reflection the conditions for its genesis. For it is not as if the "question of the world as a whole" had never been posed—was this question not the basis of Greek philosophy and science from the very beginning? The claim is rather that the world as phenomenon, as an appearing "in" that subjectivity which makes it possible, could not have been meaningful before the epoche, because the experience that forms the basis for its sense would have been wholly lacking.

This is absolutely essential to understand, if the reduction is to make any sense at all. The question of the transcendental origin of the world for Husserl becomes possible only given an *experience* which is in turn made possible only by the epoche. This means that the epoche is not simply a re-calibration of the traditional question about the world as a whole, but is itself the origin of a different mode of experience—a "transcendental experience":

> The accomplishment of the total transformation of attitude must consist in the fact that the infinity of actual and possible world-experience transforms itself into the infinity of actual and possible "transcendental experience," in which, as a first step, the world and the natural experience of it are experienced as "phenomenon."[15]

Landgrebe, in his essay on the lectures and manuscripts published in Husserliana VIII under the title *Erste Philosophie*, rightly emphasizes that in Husserl's writings "transcendental experience" is meant literally. Husserl's phenomenology represents the rejection of a division in principle between thinking and experience, where the former would reserve for itself the possibility of operating outside of the factical boundedness of the latter. If apodeictic knowledge is possible for thinking at all, it is only in the form of an experience which, in its course of self-articulation, is the event of an unfolding of an absolute evidence.[16] Thus when, in *Erste Philosophie*, Husserl follows Descartes' lead in the search for a principle *fundamentum et inconcossum*, the goal is not formulated in terms of a basic principle or element of knowledge, or even a field of apriori knowledge that would precede any and all experience, but is instead understood as an absolute evidence that is and can only be the correlate of an "absolute experience." In *Erste Philosophie*, this absolute experience was, again following the lead of Descartes, the self-experience of the ego; but unlike Descartes, what is emphasized in these lectures is not so much the truth formulated in this experience, the certainty of the judgment "I am," nor even the method that leads us to the formulation of such a judgment in its clear certainty, but again the unique *experience* which opens up, so to speak, between truth and method.

What is it that is experienced, when experience is one of the subjectivity of the world grasped after the epoche? Here too, is an experience of the world, but from the perspective of an interest in that which "occasions" this world, the subjectivity in which the world first attains its sense of an event, an occurrence (and with that its originary factical character). But the subjectivity in which the world occurs is visible only in its correlation with that which it occasions, namely the world—but the world in a unique sense, experienced now as "phenomenon." There is here a certain mutually dependent, double visibility between world and subjectivity; I cannot bring the one into focus without at the same time making the other a theme for seeing. The world as "phenomenon" is what it is only as a correlate of that subjectivity "in which" it appears; vice versa, "transcendental subjectivity" is visible only as that in which the world as phenomenon is fashioned, or achieved. Thus "transcendental experience," one could say, is a particular kind of experience of experience, where experience is experienced as an achievement. Like any achievement, its product has a double face. On the one hand, the achievement, as something in and for itself distinct from what is achieved, is a present but hidden face. To draw an analogy, one does not in a strict sense "see" the achievement of the artist in the work, only the achievement of the work itself; the work is there before us in a visibility in which the subjectivity of the artist is to be sure implicated, but not seen. The achievement I see is only that of the work, it belongs to the work more than it does to the one who created it; when an accomplishment comes to rest, the subjectivity that made it possible recedes into the background. Once the work has been achieved, the lines of correlation between artist and work are only suggested, no longer a living presence in which the subjective movement of the artist can be experienced.

On the other hand, from within the experience of the artist at work this same achievement can have another face. In its gradually crystallizing presence during the work of creation, the work of art is still within the grip of that which makes it possible, thus its "appearing" has a fundamentally different character. It is not an achievement at rest, but in motion. In motion, its givenness is no longer indifferent to the accomplishing activity of the artist, but becomes visible, a theme at all, only in a direct correlation with the activity of the artist, thus only when the subjectivity of the artist is itself experienced in the foreground. If the subjectivity were not so experienced, if it remained everywhere only implicated and never a theme of a reflection, then the work would not appear *in this sense*, namely as a given unfolding in an act. And vice versa: this accomplishing subjectivity is itself visible only when, based upon the forward movement of its accomplishment, we turn it back upon itself, making it still, turning it into something to be seen. That is, we make this life, as that which stands in correlation to the phenomenality of the givenness which it in turn makes possible, a theme of reflection.

It is in this sense that "reduction" means "phenomenalization." The result is the uncovering of a correlated, double visibility of subjectivity and world, revealed in arrested motion. And it is only at this point that Husserl can begin to formulate a conception of intentionality that is properly transcendental, for it is only within the perspective of this double visibility that intentionality is finally purified of any psychological connotations. Phenomenalization leads to an acting subject which, in order for it to be that "in which" the world appears as phenomenon, can no longer be understood as a mental act of meaning that would refer to or contain an object. Instead, subjectivity in the transcendental sense is fundamentally implicit in the basic fact of the presence of a thing in its givenness. It is the achievement of the original having of a world at all; which means that the correlation "intentional act—mode of givenness" can no longer be understood in terms of a contingent *adequatio re et intellectus*. The very contingency of adequation presupposes a horizon within which a match or a non-match can be in question. Yet this horizon itself, in which even the question of the very existence of the world as a whole can be posed as a theme for reflection, cannot be in question, but "is" only as an absolute achievement apriori.

Intentional correlation is thus an apriori structure. What is the character of this apriori? It is, as was the case with the apriori of the life-world, concrete. This means that it is not a conceptual apriori, where the suggestion would be that a specific group of concepts represents a set of irremovable, and unchangeable, conditions for experience. For Husserl, the intentional correlates are the visibility of experience itself, which in transcendental experience is given in the mode of an apriori. In transcendental experience, experience is experienced as what it is before it takes hold; the accomplishment it represents is grasped before this accomplishment has been carried through. Held still, it can only be visible as an apriori. Experience here has not settled in its own results, but has been apprehended "before" these same results, in a sense before it has become itself. But if experience is *given* in the mode of the apriori, then that means that experience is an apriori which unfolds as an experience, which is "lived through." Its invariance—always the touchstone for all senses of the apriori for Husserl—is not an expression of a claim to the irremovability or indubitability of a concept, or apriori "knowledge" (what is known about the world "prior to" experience). Instead, it is as if, in reflection, the philosophical subject "lives through" the otherwise arrested experience itself, but now with an interest in its peculiar invariance, which is present in it as something concrete, something discoverable in a twilight of arrested subjectivity, before the full emergence of the accomplishments of experience.

The analysis of the lifeworld, even if it itself is not a transcendental analysis, nevertheless serves as a guiding thread for transcendental investigation of the concrete apriori of intentional correlation. For if, as was argued above, the life-world is the original self of the world, and the sense of its original givenness is not a fixed set of meanings, but precisely a movement of meaning open to its own being-enriched and re-articulated—then it is precisely the phenomenalization of *this* sense of world that will guide the thematization of the original accomplishment of the "having of a world" as such. The basic sense of world for Husserl is the presence of the horizon of pre-givenness; and this horizon is, in turn, something explicable, or understandable, not only in a reflection on the tension between this world and its own articulation in activities such as science, but also within transcendental philosophy as the discovery of the intentional horizon of the apriori.

The great advantage of the so-called "way to phenomenology through the life-world" over the "Cartesian way" lies precisely in the importance of the theme of "horizon," as Landgrebe rightly emphasizes. The lifeworld, understood as the horizon of the questionability of the world, or the horizon of an experience of being that is not limited to a given articulation of "nature" or "world," but which is precisely the manifestation of the problematicity of meaning, orients reflection from the beginning towards the difficulties of posing the question of a transcendental apriori at all. The Cartesian way to the transcendental reduction leaves the impression that, once the being of the I has been disclosed as a field of absolutely functioning activity, a logic of the world will simply follow as the correlate of these absolute structures of subjectivity, which themselves need only be "intuited" in a pure reflective experience. In the *Crisis* we have a very different point of departure. Here the givenness of the lifeworld, as a world of *doxa*, is a givenness which is not unambiguously available as an object for intuition. Intuition here must be led by a series of questions posed with the awareness that what is given is often a complicated combination of given and non-given, of evident and non-evident, of fixed and relative. This in turn makes necessary a constant questioning into the origins of the sense of the world, and of things, in order to understand the complex play between the given and the pre-given, the visible and the invisible. Here we ask not only: how is the lifeworld given? But: how is it given, precisely when its givenness oscillates between being and appearance (*Sein und Schein*)?[17]

This style of questioning the evidence of the lifeworld also extends, within the movement of phenomenalization/reduction, to the theme of subjectivity itself. If intentional analysis is to make comprehensible the correlation between the intentional accomplishments of subjectivity and the givenness of the world, then the complications of the sense of this givenness also entail parallel complications of the sense of intentionality. Any given intentional achievement, as is the case with the complex of given and non-given which is the world, bears within itself a hidden dimension of submerged accomplishments that can only come to the fore given an attentiveness to the ambiguities inherent to its own givenness in reflection. The example of the artist above was oversimplified: the two faces of accomplishment are never so neatly separated, but always layered in such a way that one always catches a glimpse of the one in the other, like a tell-tale gesture of a parent woven into the mannerisms of a child. The result is that "intentional analysis" is to be understood as orientated around the question of origins, and not merely to what is given in a reflective, intuitive regard; and above all, for Husserl this kind of questioning into origins necessarily takes the form of an ongoing, developing field of investigation:

> Intentionality is the title which stands for the only actual and genuine way of explaining, making intelligible. To go back to the intentional origins and unities of the formation of meaning is to proceed toward a comprehension which,

> once achieved (which is of course and ideal case), would leave no meaningful question unanswered.[18]

Difficulties

In order to understand the reduction better, it is useful to consider three difficulties Husserl addresses at this point in his presentation (*Crisis*, §52).

(1) The first difficulty involves a look back to what, in the epoche, had been moved aside in order to find a "place," as I put it, for the thematization of subjective life. In the epoche, objectivity itself is set aside, in the natural sense of what is taken to be the case about the world, as the concrete givenness of things in their pre-given familiarity. But this in turn implies that we set aside a certain understanding of truth. "Objectivity" is the establishment of a certain style of the truth of things, the manner in which they are posited. Objectivity is thus not only an articulation of things in experience, but also a "claim," which is why in *Ideas I* Husserl speaks of the bracketing of the "thesis" of the natural standpoint. To be sure, the way to the reduction through the lifeworld no longer understands the positing-character of natural life to be the same thing as holding a thesis, as if somehow life itself were analogous to the formation of an argument about the nature of the world. Nevertheless, even in the *Crisis* the world is still something that takes shape in a positing activity, and thus still carries the weight of an established certainty—even if the reflecting subject is no longer unambiguously the one who "decides" that a given style should carry the weight of "truth."

If so, then is it also the case that, when this "natural" truth of things has been put aside, it has thereby been effectively replaced with another truth? Is it the case that there is here another manifestation of the being of the world which, within a newfound transcendental experience, has "weight" independently from the truth of the objective world? In short, are there two truths, one subjective, the other objective?[19]

The answer is yes and no. (a) On the one hand, there are not two truths, if by that is meant two radically different truths which have no common basis. Instead, the truth of objectivity is itself possible only as a mode of subjective life; objectivity is a value for subjectivity, its weight comprehensible only with reference to a life for which it provides an order. And if this insight is carried over into an investigation of what this would imply on a transcendental level, asking not only about the nature of the order of truth, but in turn about the basic subjective achievements that make possible a world that can be ordered at all in accordance with the idea of true being, at no point is there introduced for the "first time" a subjectivity that would be an alternative to the objectivity under investigation. The "subjective" in question is already present in objective truth, and belongs to its sense.

On the other hand, (b) the structure of objective truth is such that it fosters an obliviousness to that very subjectivity which nevertheless belongs to its essence. As a unity or style of sense, objectivity is founded on the pre-givenness of its theme, "the world." It is an articulation of the truth of things based upon this pregivenness, thus made possible by it. The result is that the universality of objective science, thus of objective truth, is inscribed within the limits of the pre-givenness of the world. If the limits of the pre-given, however, are surpassed, if they no longer operate as that which frames the question of truth and being, then the meaning of truth represented by the sciences no longer appears with the character of universality, but instead as something limited, or finite. In this sense, "truth" looks very much different when seen from the perspective of mundane experience than it does within the perspective of transcendental experience—the former can be operative only given an obliviousness to the presence of the latter, which is precisely what defines its "naturalness." Likewise, transcendental subjectivity can recognize, or take responsibility for, the meaningfulness of objective truth only by, in a sense, re-fashioning itself as a mundane subjectivity "in the world," something Husserl explores under the heading of "Einströmen." Thus even if we cannot speak of objective truth as being a wholly separate truth from subjectivity, it is nevertheless understood by Husserl to be the constitutive result of a separation within subjectivity itself—a separation we have already met under the heading of the problem of the difference between the pre-given and the given.

(2) The second possible objection Husserl takes up is the charge that, in effecting the epoche, one effectively *abandons the world.*[20] Here the answer also splits into a "yes" and a "no," though in a somewhat more subtle fashion than in the response to the first objection. (a) In an important sense, one does not, and cannot, simply abandon the world in the epoche. For the theme is precisely that subjectivity which lives the world in a manifold of different experiences. This means that world experiences themselves must remain "in effect," whether one is in the natural attitude or the philosophical attitude; they are the indispensable clues to the subjectivity that is the theme. This subjectivity, even when distinguished over and against anything worldly with respect to its manner of being, is not thereby arrived at phenomenologically in the same way as leaving one room (that of the "world") and entering another (that of "subjectivity").

Nevertheless, (b) the epoche does entail a shift that puts into question the sense in which I am "in" the world of natural life at all, even "in" the subjective activities that constitute experience. What I push against in the epoche, and in doing so separate myself by making still, is the goal of those experiences which make up natural life—that is, the goal of natural experience which is precisely the world. Experience is in and of itself directed towards the world; to turn and direct oneself to experience as experience is, at least with respect to what experience is "naturally," to disengage oneself from experience, and thus from the world as such, as that towards which experience moves. In the epoche, the world is no longer my goal, that towards which I am directed; but I am nevertheless faced with the world as the definitive theme of that "towards which" I am in fact directed: the subjectivity of experience. Thus even if the world itself is something I effectively abandon, in that I disengage myself from it, I nevertheless face it again under a different guise, as an intrinsic telos of that towards which I now direct myself in reflection.

What does this entail, disengaging oneself from the interest in the world, only to discover it as phenomenon within the perspective of a different interest? The interest in the world, the articulation of its givenness in terms of the horizon of the sense, or meaning of things (what they are, how it is that they are), is the key to the establishment of a world that we all share. We come together within the space of a shared world only given this interest. And this interest is not only limited to what things are—what this table is, how this door works—but includes an interest in one another: who we are, where we come from, how we can work together. It is an interest that fixes the world precisely in terms of our life in such a world, defined here in terms of a horizon of interests. Thus to cut oneself off from this horizon is to cut oneself off from the basis of community itself, in the sense of a shared world; it means to cut oneself off from the possibility of a shared experience.

This leads to the third difficulty (3). If so, if within the epoche one can no longer operate in terms of a shared experience of the world, then there is no common horizon within which to document what it is that can be seen; one cannot treat experience here as if it could be something shared. For that reason, the potential for fixing any "facts" of transcendental life becomes ambiguous, for a fact is something we can call on others to recognize simply by bearing witness to an experience that we share in common: all that is required is our presence within the horizon of a common world in order to give us the basis upon which we can arrive at a mutual agreement of what is "in fact the case." The facticity of the world can be fixed, it is the world itself fixed within this shared interest; but the facticity of subjectivity, possible only given a disengagement from the natural interest in the facticity of the world, is fundamentally un-shared, thus cannot be fixed as a fact. The result is that the philosopher for Husserl is radically isolated; but not as a result of an abandonment of the world in order to reflect on eternal truths independent of the facticity of everyday life. In fact, the "eternal truths" of the philosopher, once formulated, are accessible to all; as truths in a genuine sense, articulated in the language of essence, they have entered into that domain of determinability which is, in the end, an intersubjective domain. What remains necessarily outside of this domain, and which can only be taken up as a theme in the life of the individual philosophizing subject, is the *facticity of truth*. This remains, as a manifestation in factical life, a "Heraclitean stream" that can never in itself be fixed as *fact*, only as *essence*: "The fact is here, as belonging to its essence, and it is determinable only through its essence; there is no way of documenting it empirically in a sense analogous to what is done in the objective sphere through inductive experience."[21]

Thus the philosopher retreats from the world not to contemplate eternal truths, but to enter into that in which truth originally resides, where "original" means "factical."[22]

The paradox of subjectivity

These three difficulties do not in themselves lead to a reflection that would in turn demand a reformulation of the reduction, understood as phenomenalization. There is a fourth difficulty which, however, does initiate such a reflection. It was already touched on above when it was noted that our interest in the world also includes an interest in one another—which obviously highlights the fact that we encounter one another in the world not only as things, but as subjects. The sense of subjectivity is therefore not something wholly alien to natural life; on the contrary, it is something that we fix within the horizon of our life in the world, which means that it is something we share, based upon our shared interest in the way things are. This even extends to science, of course, and forms the initial empirical basis for psychology. And this means, in turn, that the sense of this subjectivity, what it "is," is part of the obviousness of the world, of what is given within its horizon of familiarity. This "obviousness" of subjectivity is precisely what leads to a paradox.

One of the common, obvious questions to ask about any subject is: "who is it?" Working late at night I hear a sound at the door, something like a knock; my reflex is to call out "who is it?" The presence of other subjects is so ingrained in the very fabric of the world, stamping all circumstances with the familiar possible of the presence of others that, even when I discover that it was only the combination of the wind and a loose latch, this exception only confirms the rule that the encounter with another subject is to be expected, if not now then later. And likewise we are forced, due to the weight of familiarity, to ask of subjectivity even after we have accomplished the reduction, "Who is it"? Who is this subject, in whose activity the world as phenomenon is constituted? One says in response "I"—but who is this "I"? The only answer we have at hand is the same one we always give, and will always give, in natural experience, because this is the only "I" we know: the persons we are, with whom we are familiar, and who are known by others. But if so, if this answer is unavoidable—and it is, which is a function of its "obviousness"—then we fall into an equally obvious paradox: how can it be that this "I," the I that I am, is both the one for whom the phenomenalized world is manifest, and a someone who is part of that same world-phenomenon? How can it be, that I am the subject constituting my own objectivity, thus the accomplishing subjectivity of my own appearance? And if the world is "world" only as the horizon of an intersubjectivity, then how can we avoid the conclusion that this so-called "transcendental intersubjectivity" is simply humanity itself—the existence of this shared factical life which unfolds within the horizon of the world? But, again, how can intersubjectivity both be *within* the horizon of the world, and that *in which* the horizon of the world becomes manifest as phenomenon?

> Universal intersubjectivity, into which all objectivity, everything that exists at all, is resolved, can obviously be nothing other than mankind; and the latter is undeniably a component part of the world. How can a component part of the world, its human subjectivity, constitute the whole world, namely, constitute it as its intentional formation, one which has always already become what it is and continues to develop, formed by the universal interconnection of intentionallly accomplishing subjectivity, while the latter, the subjects accomplishing in cooperation, are themselves only a partial formation within the total accomplishment?[23]

Husserl characterizes this paradox as a conflict between the disinterested observer one has become presumably thanks to the epoche, and "common sense" which is, and can never not be, the origin of the very sense of self. Here we have the culmination of a double movement we have been tracing throughout the *Crisis*: on the one hand, the questioning of what is unquestionable, the obviousness of the world as the pre-given horizon of experience; on the other hand, the process in which the subject becomes a problem for itself. Phenomenology as philosophy—which can have no other program than putting into question the obviousness of the world[24]—necessarily leads to the result of one becoming a problem, a paradox, for oneself.

Could one not also point to this paradox as a clue as to how to clarify the sense in which we can speak of a "transcendental experience"? Epoche and reduction lead to an experience characterized by the progressive problematization of subjectivity—which is always "my" subjectivity—whereby reflection follows a line of tension between the self with whom I am familiar, and another face of this same self which first comes to prominence as a result of phenomenalization. Transcendental experience, at least at first, can perhaps be characterized as this experience of the self as problematic.

Above I spoke of the artist as the subject of the activity of creation, and suggested that the work has a different face from within the actual movement of this activity than it has once it has become an established fact, a "finished" work. In the former there is a dependence on the subjectivity of the artist which, once the work has been completed, is hidden from view. This is analogous, I argued, to the dependence of the world as phenomenon on subjectivity. One could perhaps at this point pursue this analogy further, and say that the subject has another face to show as well. The subjectivity of the artist is not simply the face of the artist as a given empirical individual, with all the characteristics that make up the concrete person. For is it not the case that the work, as something created, is *not* dependent, in a radical sense, on the artist as a person? For what would it mean to identify its dependence on subjectivity as the dependence on this or that person—wherein would lie the dependence, on what characteristic, what pattern of behavior, what factor of personality? Is it rather not the case that the combination of *this* person and *this* work of art is something in the end contingent, that there is nothing in the one which necessarily leads to the other? With this contingency in mind, we can turn to the act of creating the work itself and recognize that the "subject," in the ordinary sense of the term (the person, the concrete "I," the one who is given as present, who belongs in this story of the event of creating a work of art), is to be distinguished from something else which is the necessary correlate of the dependency of the work on an "act" in order to "be."

This "something else" on which the work is dependent, and which is different from the concrete I, who in a way bears witness to the event, we can also call subjectivity, or "I"—but not in an unequivocal sense. At first what is important is the tension—which is none other than the tension of the paradox of subjectivity itself, but now made more precise by use of a distinction between two senses of subjectivity. This is an important feature of Husserl's argument at the end of Part IIIA of the *Crisis*: the distinction alone between the "I" as "Mensch" and "I" as "ultimately functioning accomplishing subjectivity," "Ur-Ich"—or more simply, the "I" and the "original I"—does not *resolve* the paradox. On the contrary, it *sharpens* it, and shows that the paradox is in effect irresolvable, unless one finally breaks with the idea of a simple identity of the I and the original I, and leave behind once and for all the model of the I as "the one" who performs acts, whether they be acts of a concrete self or those of a transcendentally functioning subjectivity. That is, the paradox disappears only once we have given up the need to ask the question "who is it?", and recognize instead that this "it" is itself the place where the most basic relations are set up in which any "who" is first "constituted." The original I is original precisely to the extent to which it is the origin of self, whether it is the individual I or the self as it is constituted in intersubjectivity, in the transcendental "we." It is here, on the level of the original I, that both the I as myself and the self of ourselves as an interrelated group is originally declined.[25] And here again Husserl turns to the isolation of the philosophizing subject as a *conditio sine qua non* for the development of this thematic.[26] For the sense of "origin" here is in tension not only with the "world," with "others," but also with subjectivity constituted as I; the subjectivity that is the theme of the philosopher is not only the subjectivity which is the origin of the sense of the world, but which, more importantly, is the origin of egological being itself.

And here again one can perhaps turn to the artist analogy. That subjectivity which, given a certain perspective of analysis, shows its face in the act of production is not only the origin of the work itself, but is also the origin of the artist. To understand what it is for a given person to be "the one," the one who has created this work, we ultimately need to turn away from the artist as any given "one" person and instead make thematic the subjectivity operative in the event of creation. This event, the creation of the work of art, is as much the creation of the artist as the work; and if we are to call this origin of the event something "subjective," we can do so only if we recognize it as a subjectivity which, all the while keeping for itself a certain reserve, fashions itself as the subjectivity of the artist. But that means that this face of subjectivity is again double: on the one hand, we have the ultimate functioning subjectivity as the transcendental origin; while on the other hand, we have the fashioned subjectivity as a phenomenon, itself dependent on original subjectivity and in turn a part of the world.

Nevertheless, even as it fashions itself as a "part of the world," the original I does not thereby reside in the world, but remains in tension with it. This is a tension that can be articulated, or made a theme for thinking at all, only from within the self-reflection of the philosopher—which means, again, only from within the radical isolation of the philosophizing subject:

> ...each human being "bears within himself a transcendental 'I'"—not as a real part or a stratum of his soul (which would be absurd) but rather insofar as he is the self-objectification, as exhibited through phenomenological self-reflection, of the corresponding transcendental "I." Nevertheless, every human being who carried out the epoche could certainly recognize his ultimate "I," which functions in all his human activity.[27]

To again stress: the original I is in everything that one does, but it is not everything that one is; furthermore, it is visible only from a perspective in which being is held still, thus held apart from becoming, an act thanks to which the subject reflects on its becoming as the original possibility of its being. And if my theme is my own subjectivity as a becoming—which I first pursue by making thematic the world as a flux of becoming (the "way through the lifeworld")—then I cannot identify it as myself in the manner in which I am used to, for this subject is not "who I am," but the becoming within which who I am is being decided.

A problem, the problem of oneself, thus takes shape. And it has a unique structure. For example, it is not formed in the manner of a question such as "what will become of me?" It is not set up in a way whereby which the prospects of the future frame the problem of how to understand who I am. The future is always in a sense coming from the outside, and it makes who I am problematic by comparing what I am now with what I could be. Instead, what takes shape here is a questioning in which the problem of oneself is original, primary, and out of it is coming who and what I am now, not who I will be; thus it is not as if a previously stable sense of self is made problematic, but rather that the self rediscovers an original sense of itself as a problem. What frames the question of oneself here is not on the outside, but in an unexpected dimension of the inside of subjectivity.

In this way, Husserl's reflections arrive at a different sense of subjective being than had been, or even could have been, operative in the initial formulation of the epoche. At first it appears that what one does, whatever it may be—visit a museum, listen to a concert, give a lecture—one does as "onself," as "I," and that includes performing the epoche. Thus when I recognize that the original I is precisely an "I," this too seems to fall within the purview of what I do as "myself"—it is "I" who constitutes the world, it is "I" who am the subjectivity who constitutes the sense of things. And, in addition, it is the intersubjective "we" who, together, constitute a shared world of meaning; the horizon of the accomplishments of transcendental subjectivity thus appears to be the horizon of our shared humanity. But when I in turn recognize that the original I is that which accomplishes "myself," then this "I" can no longer be simply identified with myself, but can only become my theme when—and here finding the words becomes difficult—I recognize that not everything accomplished by and in my subjectivity is done on the part of this subjectivity as "me." Without having left the fold of the movement of an accomplishing "I," this ego shows itself to be not entirely my own.

In short, what is arrived at is the idea of subjectivity as itself an accomplishment—of subjectivity. Thus we have not resolved the paradox, but have instead used the paradox to reformulate the question of subjectivity itself from within the experience of its own reflective aporia. The paradox is nevertheless productive, in that it leads to the important step of "correcting" the sense of epoche: whereas before, the epoche was understood to be the thematization of an absolutely given subjectivity, now it becomes clear that this subjectivity is itself something constituted. This means that its intrinsic, essential structures will only be revealed in a perspective opened by a deeper analysis:

> At the onset of the epoche the ego is given apodictically, but as a "mute concreteness" (*stumme Konkretion*). It must be brought to exposition, to expression, through systematic intentional "analysis" which inquires back from the world-phenomenon. In this systematic procedure one at first attains the correlation between the world and transcendental subjectivity as objectified in mankind.[28]

What concepts do we need, what language, in order to be able to bring this "mute concretion" to interpretation, to expression? The resources of genetic phenomenology as they are presented in the *Crisis* would appear to be particularly suited for the exposition of a transcendental subjectivity that is, at least at first, mute. For is not the obvious, the familiar, muteness *par excellence*? And if Husserl was successful in unlocking the theme of world-having from its invisibility in the unquestionability of the obvious, is it not at least possible that his methods could be just as successful in unlocking transcendental subjectivity from the invisibility of its own concreteness?

Husserl's *Crisis* is an unfinished work. It does not contain any such exposition of transcendental subjectivity. More, the manner in which the task of such an exposition is set out, i.e. as a problem of genetic phenomenology, or of approaching the question of subjectivity as a question of origin, can be said to lead to a fundamental ambiguity as to the nature of phenomenology as it is pursued in these texts. For is it at all clear that, if reduction is interpreted as a questioning into origins of the sort presented here, then "phenomenology" proper is not ultimately left behind for the sake of a peculiar hermeneutics that is no longer guided by the concept of phenomenon? Do not the very themes of obviousness and familiarity, sedimentation and establishment hold open the possibility that the exposition of what is implicit in the mute concreteness of transcendental subjectivity may lead reflection outside of the confines of an apriori correlation of subjectivity and world? The question, at the end of the *Crisis*, seems to be no longer the origin of the sense of the world, but the origin of the sense of subjectivity itself.[29] Can such a question be adequately posed, even answered, from within the epoche?

Notes

[1] This represents another shift of emphasis when compared to more traditional readings of the *Crisis*. So for example R. Boehm proposes three basic theses with respect to the lifeworld ("Husserls drei Thesen über die Lebenswelt," in: *Lebenswelt und Wissenschaft in der Philosophie Edmund Husserls* (1979) 23), two of which—the invisibility of the lifeworld and the failure of modern science to make the lifeworld qua ground its theme—have already been stressed above. The third, that phenomenology moves beyond science by making the lifeworld its explicit theme and therefore frees itself from it, I do not dispute, but would rather stress that the only bond phenomenology breaks with the world is that of naivete. The extent to which this "frees" phenomenology from the lifeworld, as well as whether this is a desideratum at all, will remain an open question here, though there are certainly passages in Husserl's writings of this period that argue that this is the case—for example Hua XXIX 183:13-31.

[2] Crisis 136; Hua VI 139:3-17. Cf. C 3 III 33b, where Husserl describes a stratum of "dormancy" (*Schlaf*) in the concrete I, which admits of all validities and habitualities without at the same time extinguishing them; they are not "awake," but nor are they simply gone, but remain in place: "Im konkreten Ich liegt eine Grundschicht des 'Schlafes.' [...] Alles, was für mich oder 'in' mir wiedererweckt werden kann, aber nicht wiedererweckt ist, ist im Schlaf; darunter all mein in 'Habitualität' noch fortbestehendes mir Gelten, für mich real Sein, mir bekannt Sein, mir Eigensein als Theorie, mir eigen als mir zugeeignetes Schönes, als mein Werk, meine Tätigkeit, als mein Entschluss."

[3] Crisis 137; Hua VI 140:9-20.

[4] Ideas I/Hua III,1 §§31-32.
[5] Crisis 150; Hua VI 153:3-10.
[6] Crisis 150; Hua VI 153:11-31.
[7] Crisis 137; Hua VI 140:27-33: "Perhaps it will even become manifest that the total phenomenological attitude and the epoche belonging to it are destined in essence to effect, at first, a complete personal transformation, comparable in the beginning to a religious conversion, which then, however, over and above this, bears within itself the significance of the greatest existential transformation which is assigned as a task to mankind as such."
[8] "Ich wende aber den Blick auf mich, als der ich meiner als Mensch unter Menschen in der Welt gewiß bin, und ich will diese Weltgewißheit befragen, dieses mein Sein und Leben, worin und in welchen Geltungsstufen Welt mit ihrem ganzen Inhalt, ihrer gesamten ontologischen Struktur für mich geltend wird." Hua VI 470:1-5. This is a quote from a text published as Beilage XX to §39 that is not included in David Carr's translation.
[9] "Sage ich "Welt ist meine universale Geltung," so ist das nun so zu verstehen, daß immer schon ein Universum aller Geltungen als Einheitsgeltung besteht, in denen ich positive bin, von sedimentierten und partiell in aktuellem Gang stehenden, in einem Gang auch der Neustiftung. Und diese universale Geltungssynthese ist Synthese von abzielenden Intentionen in der Bewegung der Erzielung, Habitualisierung oder Verfehlung, und so haben wir Bewegung der Akte und ihrer Sedimentierung in eins mit ihren sedimentierten Erwerben, in denen wir dieselben Akte in <einem> neuen Modus der Behaltenheit haben. [...] Dieses Geltungsganze und das Ganze als Erwerb mit der Vorzeichnung der Zukunft und in sie hineingehenden Vorgeltungen hat immer schon eine ontologische Form—die eben *der Welt*. Es ist die Welt, die meine Geltung ist und in der ich jetzt bin, der ich bin, bin als der naive "Vollzieher" dieser universalen Geltung und bin fortdauernd dasselbe Ich, als in der universalen Willentlichkeit seiend und sie im Fortführen im ontologischen Sinn erhaltend, und zwar fortführend." Hua VI 470:46-471:17 (Beilage XX).
[10] Thus the "genesis" of subjectivity is the genesis of the objectivation of the world. The explication of this genesis, which Husserl understands to be a kind of history of intentionality, is as a result a parallel history of the world, in a peculiar sense. Husserl: "Eine [...] 'konstitutive' Phänomenologie, die der Genesis, verfolgt die Geschichte, die notwendige Geschichte dieser Objektivierung und damit die Geschichte des Objekts selbst als Objektes einer möglichen Erkenntnis." Hua XI 345:16-19. De Almeida draws the distinction thus: "Man könnte sagen, um diesen Unterschied zu verdeutlichen, daß die statische Phänomenologie einen Querschnitt durch den Erlebnisstrom legt, während die genetische ihn im Längsschnitt darstellt, wodurch der Fluß und somit das Prozeshafte der Intentionalität ersichtlich wird." *Sinn und Inhalt in der genetischen Phänomenologie Husserls* (1972) 7. Cf. A. Steinbock, *Home and Beyond* (1995) 37-42, and A. Montavont, *De la passivité dans la phénoménologie de Husserl* (1999) 32-41.
[11] "Diesen ständigen universalen "Vollzug" inhibiere ich in der Epoché, und sie ermöglicht mir eine universale "Besinnung" über mein Ichsein, als das diese ganze Willentlichkeit ständig hat, hatte und haben wird, sofern es sie inszenierte und verwandelte und dabei diesen Stil des totalen Erwerbes shuf. Statt im Vollzug sein, in fortschreitender Aktivität sein, immerfort intendierend, immerfort auf das aus früheherem Intendieren und Verwirklichen fortgeltende Seiende gerichtet sein und in dem Sein der Welt fortweben, halte ich in gewisser Weise still." Hua VI 471:18-26 (Beilage XX). One could perhaps discern here the way to a possible solution to the difficulty raised by H. Fein commenting on these same passages (*Genesis und Geltung in E. Husserls Phänomenologie* (1970) 39): in what sense can we still speak of a progression, in stages or levels (*Stufe*), of the clarification

of the achievements of transcendental subjectivity, if our point of departure is this self-grounding Vollzug? Fein: "Die >>Evidenz<< im Sinne des >>ich sehe es klar<< bliebe als einziges Kriterium, ohne dass auf die Möglichkeit, in welchem Umfang ein solches Axiom Allgemeingültigkeit haben kann, reflektiert worden wäre oder auf die Frage, wieso es beim transzendental leistenden Subjekt überhaupt zu >>Stufen<< von Klarheit kommen kann und nicht vielmehr eine reine stufenlose Erkenntnis im Sinne einer >>visio beatifica<< vorliegt." A suggestion: still-life is a constructed distance of the I from itself, within itself.

[12] Hua VI 471:35-36 (Beilage XX).

[13] It is because of this that, in an important manuscript on passivity from 1931-1933, Husserl can describe the "waking day" as a kind of reduction: the period of wakefulness that makes up the day is a kind of natural holding still of something which, in its essence, is never held still—the life of subjectivity itself. The day holds still what I am, in that it holds me in wakefulness against my past—that is, against the presence of memory. Husserl: "Ferner, es muss dann ausdrücklich gemacht bzw. einsichtig gemacht werden, dass eine Art Reduktion es ist, die eine Wachperiode ergibt, eben in einer Betrachtung, die auf die in die anderen Perioden hineinreichenden Wiedererinnerungen keine Rücksicht nimmt [...]." Furthermore, waking life also represents a kind of natural reduction precisely in its provisoriness, or in its only apparent holding of the past in suspension—because the past, in the form of habituality, is "irreducible," that is, it is not something from which I can "actually" cut myself off: "Es zerstört doch mein personales jetzt in dieser Wachheit Sein, wenn ich die Habitualität meiner konkreten Zweckordnungen etc., die doch nicht von heute sind, ausser Geltung setzten wollte." D14 3a. This tension between my "day" and what I ultimately am, I would argue, prefigures the more "unnatural" tension between the invisibility of the lived world in naïve belief and the "day" of phenomenological reduction.

[14] Crisis 152; Hua VI 155:23.

[15] Crisis 153; Hua VI 156:15-19.

[16] Landgrebe: "Wenn nämlich alle Selbstgegebenheit letzten Endes auf absolute Erfahrung zurückweist, so ist damit nichts anderes als das positivistische Prinzip der 'normativen Kraft des Faktischen' ausgesprochen, mit dem jede Möglichkeit der Unterordnung des faktischen Gegebenen unter die Apodiktizität ewiger Wahrheiten aufgehoben ist. Der verborgene Sinn des Positivismus als eines Protestes gegen die Unterdrückung der Faktizität durch das metaphysische Denken wird hierdurch mit aller Deutlichkeit sichtbar..." *Der Weg der Phänomenologie* (1963) 172-173.

[17] Crisis 156; Hua VI 159:4-5.

[18] Crisis 168; Hua VI 171:12-17.

[19] Crisis 175; Hua VI 179:5-6.

[20] Crisis 176; Hua VI 179:37-39.

[21] Crisis 178; Hua VI 182:6-9: "Das Faktum ist hier als das seines Wesens und nur durch sein Wesen bestimmbar und in keiner Weise in analogem Sinne wie in der Objektivität durch eine induktive Empirie empirisch zu dokumentieren."

[22] Thus Husserl appropriates the Augustinian motto, "in te redi, in interiore homine habitat veritas," where the emphasis on "habitat" is just as radical as it is in Augustine: for in both philosophers the central issue is the openness of life for truth, thus for a movement in which facticity becomes refashioned in self-responsibility—not for the truth, but for itself. Landgrebe rightly emphasizes that, if we are to talk about transcendental subjectivity as 'absolute,' it can only be something similar: "So stellt sich heraus, daß die 'absolute Erfahrung,' auf die sich alle Verantwortung und Rechtfertigung des Lebens gründet, eine geschichtliche Erfahrung ist. Ihre Absolutheit und Endgültigkeit beruht nicht auf einer

Erkenntnis einer an sich seienden Wahrheit, nicht auf dem Ergreifen oder Überwältigwerden von einer 'ewigen Wahrheit,' sondern sie ist absolut im Sinne des Gestelltseins vor einer unübersteigliche und nur hinzunehmende Faktizität." *Der Weg der Phänomenologie* (1963) 188.

[23] Crisis 179; Hua VI 183:15-24.

[24] Husserl: "Die Welt ist das einzige Universum vorgegebener Selbstverständlichkeiten. Von vornherein lebt der Phänomenologie in der Paradoxie, das Selbstverständliche als fraglich, als rätselhaft ansehen zu müssen und hinfort kein anderes wissenschaftliches Thema haben zu können als dieses: die universale Selbstverständlichkeit des Seins der Welt—für ihn das Größte aller Rätsel—in eine Verständlichkeit zu wandeln." Crisis 180; Hua VI 183:38-184:6.

[25] Cf. D8 6b: "Äusserlich müssen sich die Egos objektiviert zeigen als Glieder der Welt, als Subjekte in der Welt; innerlich bedarf es keiner 'Individuation'; es ist absolut und als absolutes Sein die Voraussetzung selbst für alle reinen Möglichkeiten."

[26] Crisis 184-185; Hua VI 187:31-188:2.

[27] Crisis186; Hua VI 190:3-10.

[28] Crisis 187; Hua VI 191:9-16.

[29] Again, I will not be dealing with Part IIIB, or the "way into phenomenology" through psychology. But there the story is similar: the paradoxical character of subjectivity, which results in a crisis of psychology, leads necessarily to the phenomenological elaboration of subjectivity. That this crisis of the sense of subjectivity in psychology is a symptom of the crisis of reason itself is more clearly grasped, however, in the way through the lifeworld, thus the focus on Part IIIA here. Still, it is clear that for Husserl the story is not complete without a consideration of the role of the problem of psychology in modern philosophy.

Conclusion

Summary of the interpretation

Husserl's *Crisis* has been interpreted above as an introduction to phenomenological philosophy that takes as its point of departure a reflection on science. Husserl's is a reflection of the type that, in its development, or maturation, necessarily takes the form of a critique, in the attempt to articulate the fundamental gestures or thought movements of science as a mode of human understanding. As a critique, however, Husserl's reflections are intended to be more than the analysis of the forms and structures of scientific discourse; they are meant to explore the possibility for a certain kind of reflective experience, one that will at the same time point to the necessity for a specifically phenomenological reflection and its accompanying method.

Yet Husserl's goal of demonstrating the necessity for phenomenology is presented in the *Crisis* as inseparable from another goal, that of re-awakening a sense of the value and purpose, as well as responsibility inherent to rational life and culture. Such a re-awakening is to take the form of a re-experiencing of science or, better, of the *idea* of science. This new experience of science takes shape in a philosophical reflection thanks to which the compelling force of the gestures or thought movements that constitute the essence of science is re-discovered from within. It is thus in the fulfillment of a renewal of the scientific experience that phenomenology will be shown to be indispensable, thereby legitimating its claim to the status of a genuine philosophy, and not merely one science among others.

A fundamental assumption of Husserl's in the *Crisis* is that science has become an organic part of the way that we think and understand, even exist. This implies that "for us," science cannot be approached meaningfully from the "outside," as one way of thinking among others. It cannot be put into question without at the same time our humanity being put into question. A reflection on science for Husserl thus represents a self-reflection of modern humanity, as does its critique a self-critique of the same. And if it is to be complete, therefore, such a critique of science can only be an immanent critique, one that takes as its point of departure the basic supposition that we *accept* what science is and tells us, no matter how critical we may otherwise be of particular, contingent aspects or trends that have become prominent in the history of the scientific project. Thus what, in many texts meant for inclusion in the *Crisis*, begins as a full throttle polemical assault on "objectivism" as the prevailing ideology of modern scientific culture since the 19th century, is often left behind in favor of a more complex attempt to establish the basis for a critique within scientifically oriented thinking, taken as a teleologically unified set of traditions. The result is that the strategy Husserl develops is not only for the purpose of exposing the errors of objectivism, but to investigate the unity and coherence of the inner teleology of scientific culture as a living tradition, or as the dynamic flow of the self-realization of rational life. Objectivism remains the lightning rod of contention throughout; but the argument in the end is that rationalism can be freed from the distortions of its objectivist manifestation only by way of establishing, through reflection, a consciousness of the teleological sense-unity of scientific understanding.

The possibility for such an immanent critique of a mode of understanding like science lies for Husserl in the uncovering of the evidential sources for its validity, for what makes it "believable." Such sources of evidence and belief are not on the order of "objective facts"—that would only beg the question of the legitimacy of objectivism—but belong to the essential structures of "subjective experience." This means that the "claims" made by science, about what is meaningful and not meaningful, about what is worth thinking and what thinking is worth, are approached by Husserl in a very particular fashion. They are not evaluated in the ordinary sense of evaluating a claim, but are instead taken up in reflection as potential claims or positions that, so to speak, seek to make a potential claim on the subject, or to call on the subject to speak on their behalf. To be genuinely "inside" a claim is to be, in a direct sense, its "subject," the "I" who stands for it and speaks for it. This "immanent" approach of Husserl's reflections sets the *Crisis* apart as a defense of reason, for it does not really seek to convince the reader of the legitimacy of science by elaborating the reasons why science must be accepted as valid, or authoritative. It is a book that, as Husserl himself indicates towards the end of Part I, is written for those who are convinced already, those who "believe," and who are seeking to understand the meaning of their convictions in a world where such convictions have become difficult to comprehend. It is a book, it could be said, for those believers in reason who are searching for their voice, for a way to speak on behalf of an idea that has not yet learned, even after all of these centuries, to speak for itself.

This theme of *belief* or *motivation* is of central importance for the entire work. All thinking, all concepts and intellectual-spiritual activity for Husserl are motivated; there is no thinking without resource and inspiration.[1] Husserl's reflections in particular are motivated by a need to understand science, from within, or in terms of the possibility of being the one responsible for its articulation, and with that its defense. This need takes the form of the need to react against the obscurity and incomprehensibility of reason characteristic of modernity. The need to understand science in light of its incomprehensibility is, however, historical, and has a complicated genesis. It is ultimately founded in the originary motivation that gives rise to science itself, or the universal motivation for understanding in general, which can be described as a need on another, more universal level: namely, the basic human *need for a world*, for a surrounding spiritual environment in which human activity can unfold as a patterned, ordered movement of accomplishment. Science is for Husserl a spiritual accomplishment that responds to this fundamental human need for an understanding of, and thus an orientation within, a surrounding context for human activity and life. What gives the motivation for a reflection on science its complicated structure is the fact that, according to Husserl, in the current spiritual-historical situation science no longer responds to this basic human need, thus it no longer fulfills its original spiritual function and purpose. Science, however successful it may be, even however much we may "believe" in it, has developed in such a way that the theoretical accomplishments of science in the articulation of the world have lost their spiritual force, they no longer "say anything to us," and have become, in that sense, "meaningless." This means that any attempt to speak for science, which can no longer speak for itself, is equivalent to searching for the meaningfulness of science, understood as rooted in the motivational nexus of subjective life.

The apparent meaninglessness of science is not due to a lack of content as such, only a lack, so to speak, of spiritual force. Science has a great deal to say, but it is unable to speak in such a way that the importance of what it says, the significance of its truth, is sufficiently clear, even to those who are open to it. The result is that the world has itself become unclear, precisely in the form it has taken as something which has been articulated, or understood, by science. Its very evidence, secured by science, fails to compel; truth itself appears to be "false," leading to the malaise of skepticism within scientific culture, or to what other critics of western civilization would call nihilism. This leads to a paradoxical situation—the successful understanding of the world, of its articulation in cessful understanding of the world, of its articulation in concepts, strikes most as empty, as not addressing what an understanding of things needs to address. This means that the very way that we comprehend things has become "incomprehensible," and that the world we inhabit thanks to this comprehension, and the orientation it establishes for us, has become "uninhabitable."

In sum, Husserl's reflection on science attempts to find a voice for scientific understanding, one with which the believer in reason could speak on behalf of science with respect to what it means to be a rational human being, or to live a life in reason. What motivates such a reflection is the historical need to re-connect the accomplishments of science to their original motivation, the universal need for a world, in such a way that the accomplishments of scientific consciousness can be shown to be not only of value to humanity, but charged with sufficient spiritual force to address the "burning questions" of our times.

Such a reflection cannot take as its object "science" in the sense of an abstract unity of propositions and theories, but must orient itself to science in the concrete, or to science as an intrinsic part of the manner in which the world is experienced by modern humanity. The "meaning" of science and the life experience of modern civilization are, for Husserl, inseparable. In particular, Husserl understood that an immanent reflection can access the claim of science only if it orients itself with respect to that dimension of life thanks to which such a claim can be made on the concrete, individual subject at all. For the claim of science is not an ordinary claim, but a claim that is rooted in, and shaped by, fundamental human motivations. Human life is in need of a spiritual surrounding world; science responds to this need by making a claim with respect to what is meaningful and not meaningful, thus with respect to the kind of world in which human existence can have a place. "Reason," rational scientific consciousness in the concrete, articulates the claim that genuinely human life is rational life, that human understanding is rational understanding, and that the world is a place for humans only to the extent to which it can be understood, and experienced, in light of its rational comprehensibility. Even when formulated on this level of generality, it soon becomes clear that such a claim can be made only in the form of an *historical task*, that is, the historical task of thinking within the horizon of the idea of an understanding that articulates the world solely within the limits of the rational. As a task projective of a way of life, the claim becomes in this way a concrete phenomenon, which is possible thanks to its historicity, or the human capacity to engage tasks and claims on the plane of historical existence.

What is characteristic about historical tasks is that they are never assigned once and for all. Likewise, claims about what is meaningful and not meaningful in whose name such tasks are established, or that form the sense-core of such tasks, are never fixed once and for all. This implies that the meaning that animates the task of science "means" differently in different historical periods, though for Husserl there is a historical-teleological unity that determines a total sense or meaning (*Totalsinn*) of which each historical manifestation represents a finite, limited moment. Nevertheless, the historical character of this totality implies that its sense is something that is constantly re-worked and renewed in different historical articulations of what is called for or demanded in order to fulfill the ideal, or the promise of the concrete realization of rational understanding. Science, as historical, is a phenomenon of re-discovery, which means the renewal of the self-understanding of what it is to be rational, even if at the same time the claim of science casts itself as something universal, true for "all times." In fact, the universality of the claim of science is not contradicted by its historicity; the argument is that the historical is precisely what enables the universality of science to be present to consciousness at all, as formative of a concrete horizon of subjective life. Phenomenological philosophy itself is understood by Husserl as an expression of this universality of the idea of science, one that recognizes the dependency of any sense of the universal on the historicity of thinking.

In Part II of the *Crisis*, this conception of a universality that is present thanks to an historical process of establishment—of its articulation and re-articulation—prefigures the phenomenological theme of constituting subjectivity. The "need for a world" is addressed through the historical accomplishment of projecting a path of self-realization, where such a projection represents the movement of a consciousness that shapes or forms itself as a directedness towards the idea of a rational comprehension of the world. Subjectivity shapes itself as the horizon of an encounter with things. Subjectivity thus not only shapes the givenness of the given, but it also shapes the sense or givenness of its own potentiality or possibility as subjectivity as well. Consciousness in this way not only has a sense for the world, but a sense of this "having a sense for the world" itself, one that is determined by a characteristic, historically relative style. One could argue that the very fact that the style of world-having is recognizable as "historical" simply means that it can take the form of a task, where what is projected is a new possibility to be realized, and not merely the possibility of indefinitely projecting the given, already present possibility into the future. The subjective act of instituting an historical task, or projecting a possibility for itself in an historical mode, shapes the very possibility of its own future formation of sense.

This projected path of self-realization—the historical "task"—is for Husserl something that is uniquely addressed to "persons," understood not simply as individuals but communities and generations as well. The formation of a horizon of sense, of historical tasks, is not an anonymous process, but is bound up with the phenomenon of the personality, which for Husserl is distinguished by the structure of self-consciousness. The basis for the historical reality of the person, of the one who has a history and who becomes "historically," is the unfolding of a consciousness of the horizon of sense and significance in terms of the determinations of selfhood, of "who we are." However anonymous other aspects of human life and subjectivity may be for Husserl, his argument in the end is that the sense-unity of life has a fundamental origin in self-constituting subjectivity, one that originally establishes itself concretely as a self-consciousness. Nevertheless, no self-consciousness is simply given in the form of a complete, established concrete person, above all one who could stand as the subject who speaks on behalf of the claim of science. Again, the claim of science can take shape only within a subjectivity that has fashioned its own emergence in life as something historical, which means as a phenomenon of renewal. This implies a systemic incompleteness at the heart of self-consciousness, one that is arguably implicit even in what Husserl calls the "final establishment" (*Endstiftung*) of the task of reason in phenomenological philosophy.

The insight into the systemic incompleteness of self-consciousness is one reason for Husserl's interest in the Renaissance, above all as it is manifest in the figure of Galileo. In Galilean science, a "new" idea of humanity and thinking finds its origin; yet it is an origin that is dependent upon its position within an historical movement, and in two senses. First, (1) the idea of humanity that arises out of the Renaissance is the result a self-conscious act of rediscovery and re-origination of the Greek origin of science. Thus the task of science in its "Galilean" form is not only to project a certain mode of possible understanding, but to re-discover at the same time the Greek idea of understanding that stands as the "historical" origin of the Galilean "origin" itself. That an "origin" is not something uniquely positioned and encountered in the span of historical time, but occurs in multiple encounters throughout history, is a key thesis of the *Crisis*. (2) The manner in which Galilean science was instituted historically necessitates in turn another rediscovery, on the part of those who "come after" Galileo, of the very origin that Galileo himself represents. The "origin" or "beginning" represented by Galileo is not only the first emergence "in history" of a particular style of self-awareness, but also a "beginning" that itself ultimately becomes a problem, a question to be addressed. The question will be pursued from within historical consciousness, to be sure, but it will address the need to "discover" the origin of sense not of Renaissance science, but contemporary science. Thus an analysis of the Galilean origin not only looks behind, to the Greeks, but forward as well, to the situation, and questions, of those who must re-invent the idea in the wake of the Galilean establishment (*Stiftung*). Historical questions are here also contemporary questions, thus require a zigzag mode of analysis.

This means that the Renaissance is thus not only a renewal, but a beginning that, as "another" beginning or origin, reflects the historical character of all origin as such. Husserl's reflections on historical origins yields a conception of historical awareness that does not limit it to the retrieval and refashioning of knowledge from a distant past, but sheds light on the manner in which the discovery of possibilities that belong to the historical present can occur thanks to the existence of the past "in" the present, in the form of a tradition. Tradition is a kind of concrete, passive historical memory that sets the potential for the present into relief, thereby giving it form and texture. For this reason, a critical reflection on the sense unity of the historical task of science requires a reflection on the constitutive role of tradition: there are possibilities, real possibilities, that are set into motion thanks to tradition itself, even in the absence of a discovery of a forgotten relic or piece of knowledge from the past. This is where Husserl's reflections, especially in the *Origin of Geometry*, take on a truly original character: the argument is that there is a decisive role played not only by the explicit historical awareness of a thinker like Galileo, for example his reading of ancient geometrical and philosophical texts, but also by what has become, thanks to traditionalization, "obvious" to a given historical period. The *Origin* even takes this to the limit of situating the entire reflection on the origin of a thinking oriented towards ideality within the scope of what was "obvious" not to Galileo, but to "us," those who inhabit the historical present. What is available thanks to history belongs to the very fabric of lived experience, which is not only historical in the sense of having an objective origin in the past, but also in the sense of the living subjective evidence of the very presence of history itself, characteristic of a forward-moving progression of understanding. Here again the theme of a constituting consciousness is prefigured: the reflective availability of the movement of subjectivity, of its inner development, will be the guiding thread that leads to the possibility of accessing the phenomenological flow of consciousness in its purity. The theme of "historicity" in texts such as the *Origin* dovetails in this way with the theme of the self-accessibility of transcendental consciousness.

The emphasis on the theme of "obviousness" in historical reflections, ubiquitous in the texts of this period, allows Husserl to re-cast the theme of the lifeworld, or the subject-relative world, in a way that would not be possible from within the ordinary perspective of science. For the scientist, the lifeworld is subject-relative in the sense that its evidential character is undermined from within by the perspectives inherent to the interests, styles, and ways of thinking that belong to individual human subjects. Given what these subjects are, the world "looks" a particular way from their perspective, and this is often not compatible with the way that the world "is" in fact. But Husserl's presentation of the theme of the lived world, emphasizing the character of the lifeworld as what is "obvious," effectively circumvents a basic presupposition to the approach to the subject-relative in science, but without directly contradicting it—the association of the lifeworld with a particular understanding of what it is to be a subject, namely as a concrete being burdened with particular interests and unavoidably contingent and unreliable patterns of thought. The emphasis on the obvious evades this gesture at a stroke, for there is something about obviousness that is immune to the interpretation given to it by science, or any interpretation for that matter. For when something is genuinely "obvious," there is no need to trace it anywhere, much less to a particular set of interests or perspectives that belong to a concrete subject; the obvious is, almost by definition, always overlooked. It still remains something "subjective," but now almost by default, since there is also no need to establish that what is "obvious" is also the case "in fact," or "objectively." The sense of obviousness thus vacillates between the subjective and the objective; it belongs both to subjectivity and to the world, and can be arguably approached as a unitary phenomenon: the subjectivity of the world.

This opening provided by the ambiguity of the distinction between "world" and "subjectivity" in obviousness is all that Husserl needs to employ the theme of the lifeworld as a "way into phenomenology." Insofar as the sense of subjectivity remains ambiguous, untouched by the decisions with respect to the interaction between subjective structures and the objectivity of the world that have been instituted by science, the lifeworld can be used to frame the "problem of subjectivity" in a way that will lead to phenomenology proper. This is already in evidence in Husserl's presentation of the notion of the epoche in terms of setting aside a portion of time for philosophical activity, or what he calls the "vocational time" (*Berufszeit*) of philosophical reflection. What is on the one hand a radical shift away from the natural attitude, the holding in abeyance the very activity of consciousness in order to establish the transcendental perspective itself, is nevertheless on the other hand an activity that is situated within the temporalization of the lifeworld. And in order to be situated in this manner, it must obey the rules for such situatedness—the transcendental spectator must have "time" set aside for reflection; life must establish itself and "settle" in a certain manner in order for the possibility of transcendental insight to be encountered as a concrete possibility, "in" the world of subjective evidence.

And when Husserl's presentation of the phenomenological reduction reaches what is perhaps the most intractable difficulty of his philosophy, the paradox of a subjectivity that is at once world-constituting and a part of the world constituted, the function of the theme of the lifeworld in setting up the reflection shows itself to be an important resource in elaborating the philosophical implications of this paradox. It is not, however, enough of a resource to evade the paradox altogether; what it does do, is that it enables one to recognize that the naiveté of the natural attitude conceals a paradox, and that this concealment belongs among the structures of the lifeworld itself. The very concreteness of transcendentally functioning subjectivity obscures its paradoxical character; what the lifeworld as a philosophical theme allows Husserl to accomplish is a clarification of the nature of this obscurity, even if it is not enough to dispel it. Such a clarification is not something Husserl was in a position to do given the instruments available to him in the *Ideas I*, for example. The nature of this obscurity is that transcendental subjectivity, even taking the form of a developing self-consciousness, can function in a state in which it is not clear about its own sense, that it does not even encounter its transcendental function as a problem, at least not originally. For Husserl, in fact, it is only in the development of modern philosophy that the subject becomes a problem and only in phenomenology that the implications of this problem are fully explored.

Husserl's *Crisis* is an unfinished work, and it would of course make little sense to demand completeness. Much more can be said, of course, in particular with respect to how the *Crisis* relates to Husserl's earlier work. What I have tried to do above is to lay the groundwork for such a project by providing the outlines of an interpretation of this text and related manuscripts. What I would like to offer, in conclusion, are some remarks with respect to what I believe to be the philosophical potential of the *Crisis*. This can be best carried out by again stressing the theme of *Besinnung*, as well as two other concepts that are arguably of central importance for the *Crisis*, but which nevertheless receive very little elaboration: "flowing" (*Einströmen*) and "inwardness" (*Innerlichkeit*).

Some questions: *Besinnung*, *Einströmen*, and *Innerlichkeit*

If the goal of phenomenology is the radical clarification of the essential structures of subjectivity, and if the way to phenomenology is through the lifeworld, then the clarifications achieved by phenomenology will follow the contours of an investigation of the various aspects of subjective world-experience, including the spheres of nature and community. The outline of this path, as well as an alternative path through psychology, is the topic of Part III of the *Crisis*. But it is also in Part III that the unity of the work seems to falter, and not only because of the fact that it remains an unfinished work. It is also a question of design: in phenomenological reflection proper, at least as it is presented here, there is a sense in which Husserl leaves reflection in the sense of *Besinnung* behind, and re-engages the kind of reflection that he had presented in earlier methodological texts such as *Ideas I*. The problem of the clarification of subjectivity no longer, at least explicitly, represents a task of becoming conscious or aware of a sense or meaning embodied in a tradition, or an historical task. The aim of phenomenology proper is to grasp, in a reflective consciousness of another type than *Besinnung*, the eidetic structures of lived experience that form the constitutive patterns of the comprehensibility of things.

There may be good reasons for this apparent lack of unity. Perhaps the clarification of subjectivity, especially in light of its transcendental functioning, in fact requires a different kind of reflection than one that would engage the possibility of a sense or a meaning, or a claim, as in the idea of *Besinnung* as an immanent critique of the claim of science that was presented above. Phenomenological method, it could be argued, depends on the capacity of consciousness to discover within its own concrete flowing the possibility of a kind of non-mundane self-manifestation, or visibility; for in naïve, natural life, the transcendental functioning of consciousness is not manifest, or thematic. In the *Ideas*, the potential for such self-manifestation is presented in terms of a reflection in which consciousness has modified itself, or its empirical flow, in such a way that its transcendental dimension becomes a theme for the reflecting, philosophical ego.[2] One could argue that this emphasis on modification, so important to Husserl's work after the *Ideas*, amounts to an innovation on the theory of inner perception (not the first in Husserl's career), and one that is not necessarily free from the difficulties of the conception of a subject grasping itself in a reflective act.

Reflection as *Besinnung* would not replace the dependency of phenomenological method on some kind of reflection that resembles inner perception; but it would, significantly, relieve the latter of the role of introducing the theme of transcendental consciousness on its own. Thus one could speculate that Husserl's criticism of *Ideas I* in the *Crisis*, and his apparent rejection of the so-called "Cartesian Way," should really be taken as only a limited rejection.[3] The point would be that Cartesian reflection alone, with its own resources, cannot lead to anything but an abstract notion of the transcendental ego; it has neither the evidential resources or motivational framework that would lead the reflecting subject to a clear insight into its own transcendental functioning. Cartesian reflection thus requires a supplement; but the point is not that Cartesian reflection is thereby dispensable. It may, in fact, be indispensable, even for a phenomenological philosophy that, thanks to the supplement of *Besinnung*, has found its way to the questioning into the ambiguity of subject and world that will ultimately lead to the transcendental thematic, but which cannot, or at least there is no indication that it could, transform itself into a fully developed transcendental-phenomenological method. Other resources, intuitive resources, are needed to secure the manifestation of transcendental consciousness.

This admittedly open question in Part III of the *Crisis* would seem to shift the center of gravity away from the problem of the crisis of the sciences to the problems inherent to securing a basis for phenomenological method. It may be the case that the development of a *Besinnung* with respect to the meaning of science provides a fruitful point of departure for phenomenological analysis, but does that mean that phenomenological analysis represents a further development or maturation of *Besinnung*? That is, does phenomenology really address the fundamental motivation behind the *Crisis* as a whole, namely, the historical necessity of addressing the questions that arise from the experience of the crisis of reason? This is perhaps the most important, but also the most difficult question one faces when reading the *Crisis*. For it is not only the question of how it is that philosophy is to have an impact on the cultural life of Europe in the face of the nihilism of modernity, but specifically the kind of role that philosophy could possibly assume in the form of a "transcendental phenomenology." What kind of contribution is such a philosophy supposed to represent—what impact, spiritual impact, is such a project really supposed to have?

This question becomes all the more pressing when it is recognized that, for Husserl, transcendental philosophy represents a mode of disengagement from things, not in the sense of a complete separation from the interests of life, but in the sense of a focused engagement with phenomenality at the expense of all other interests. Phenomenology holds life at arm's length, precisely in order to see things that life ordinarily does not allow to be seen; and what it holds at bay is not trivial, but essential, namely the forward movement of life, the very sense it has of its own "living." Yet this circuit of the flow of lived experience is broken in a way that nevertheless keeps it in place, in order to catch a glimpse of what occurs, but remains latent, thanks to this forward movement itself. But with its radicalized focus, or shift of attitude, such a mode of engagement is still alien not only to the interests of "natural" life, but its very style or mode of existence. This is what makes the motivational basis of phenomenology such a complex issue—and, in turn, its general significance for life as well.

In several texts from the period of the *Crisis*, both this disengagement, and what is revealed thanks to this disengagement, are presented as possible sources for the enrichment of sense or meaning even in natural life. This is one of the central themes of Husserl's philosophy since the *Logical Investigations*: the reflection on meaning, the thematization of sense, has an impact on how meaning orders and shapes lived experience. Reflection and the clarity that it achieves spawns new patterns and structures within the sense unities or complexes of meaning that order and direct all experience. Thus for Husserl, to make the meaning of things a theme has an immediate impact on the meaning of things; a sense of something cannot be articulated without shaping it at the same time, and that even includes the sense of sense, or the "meaning" of meaning: having made sense a theme for reflection, we make sense of things in a different way than before, we are no longer "naïve" about what it is to "mean."

This was already the point of defending a quasi-Platonic conception of the ideality of meaning in the *Logical Investigations*, but the same is true for the thematization of the constitution of sense-determined objectivity and lived experience in transcendental subjectivity: the dimension of sense, not only in its ideality but in its subjectivity, once articulated, *enriches* the sense of the experience that was initially the object of reflection. With ideality, the argument turned on the manner in which the rational foundations of sense and meaning can be fixed as ideal relations or structures, which then in turn serve to orient a consciousness as it makes its way around the world in terms of the order of sense. Knowledge of the "truth," as it were, about the essence of meaning is meant to play a significant role in the self-understanding of a scientific subject who regularly employs the tools and techniques of logical and rational argumentation in an investigation of nature. But things are not so straightforward with transcendental subjectivity: at best, such knowledge could lead to a reform of psychology, but the value of that is clearly more limited for science as a whole when compared to the clarification of logic. Is there another sphere in which transcendental philosophy would make a unique contribution?

Husserl deeply believes that the sphere that transcendental philosophy would enrich is life itself: all sense-unities are universally susceptible for transcendental enrichment. The mechanism whereby transcendental philosophy enriches the sense unities of natural life is designated in Husserl's later writings by the term "*Einströmen*," which describes the manner in which the transcendental dimension, or the consciousness of the subjectivity of meaning on a transcendental register, "flows into" natural life, enriching its content, above all the content of the sense of "subjectivity" itself.[4]

On one level, *Einströmen* is supposed to result in the rejuvenation of psychology as a science of the subject, one that would no longer attempt to reduce the empirical subject to a set of mathematically manipulable formulae (thus "naturalize" it in that sense), but which would recognize, and take as its guiding idea, the irreducible dimension of the transcendental. Such a psychology, to invoke the legacy of Brentano, would be a "psychology without the soul," where the reflective articulation of transcendental subjectivity would replace the traditional special metaphysics of the soul, or any positivist versions of psychology still plagued with metaphysical ambiguities. And also in line with Brentano here is the idea of a "descriptive psychology" (now called phenomenology) capable of playing a distinctly philosophical role, in part due to its methodological independence from the natural sciences, but above all thanks to a newfound sense for the explicative power of description. But for the Husserl in the *Crisis* texts, this is no longer simply a question of method, but is grounded in a deep reflection on the sense of subjectivity that results from its becoming aware of itself in a multi-dimensional manner—as concrete, human ego, or "man in the world"; as historical subject, "called" by a historical task; and finally as a transcendentally functioning ego. In sum, whatever one thinks about Husserl's claims about the force of the idea of transcendental philosophy, it is still the case that, because of its comprehensiveness and ambition, Husserl's philosophy represents a potentially powerful approach to taking stock of the impact on self-consciousness that has resulted from the explosion of different conceptions of subjectivity that has characterized modern philosophy. For one of the fundamental questions facing contemporary philosophy in the wake of the legacy of the 20th century is: what would a self-consciousness be, for a self that has learned to understand itself in all of these manifolds of sense?

But on another level, and thanks to the way into phenomenology through the lifeworld, *Einströmen* modifies the sense-content of the surrounding world as such, or the sense of the world *as world.*[5] But it does this not only because the "results" of transcendental investigation contribute to the form and complexity of the experienced world, nor does transcendental investigation rejuvenate the experience of the world, but because the possibility of transcendental philosophy, and with that the possibility of philosophy as such, shapes the very potentiality of life to project itself into its own lived horizon. What changes thanks to the transcendental elaboration of the sense of subjectivity is the manner in which concrete subjects function as subjects. One not only has a different understanding of what it is to be a subject, in the sense of a theoretical advancement; but the style of subjective life itself is modified, established in accordance with a different sense of the being of the subjective, one that no longer filters itself through the ambiguity between subject and world.

This leads to what I would argue to be one of the most interesting, if undeveloped concepts found in Husserl's later writings: the idea of a "transcendental inwardness" (*transzendentale Innerlichkeit*).[6] What is interesting about this notion is that it does not belong to the transcendental spectator as such, but results from the manner in which a phenomenological reflectivity embeds itself in natural life, though without at the same time becoming absorbed in it. That it is, it seems to be not so much the inwardness of a reflecting subject as it is the resulting inwardness that takes hold in a concrete subjective existence in the wake of reflection. It is the result of what the very event of philosophy has made possible, or the kind of awareness that results from having taken the philosophical course of reflection in thinking. Thus in a sense it is independent of the specific results of philosophy, which may or may not take the form of a genuine "philosophical knowledge," and instead has to do with a developed sense for or insight into what the possession of such knowledge "means," what it says about who we are.

It is an inwardness which, one could say, is already a feature of our concrete life as the result of being in possession not only of a body of knowledge about the world, but a tradition of philosophy charged with the task of understanding the significance of knowledge as such. It is one thing, for example, to have arrived at the knowledge that the universe is infinite; it is another thing to arrive at an insight into what it means to be a creature who understands that the universe is infinite. The same is true of philosophical reflection: it is one thing to show that philosophy is possible; it is quite another to understand what kind of philosophical subject, or what kind of philosophical life, is the result. Thus the development of a "transcendental inwardness," one could say, is precisely meant in the sense of a rediscovery of a dimension that is in place thanks to the philosophical moment embedded in the surrounding cultural world, where this moment is reestablished on a "transcendental" register. In short, all of life is enriched for Husserl through the rejuvenation of philosophy, and its subsequent reformation into a transcendental philosophy.

What does such an "inwardness" amount to? (What does "philosophy" amount to?) First, for Husserl it represents a heightened sensitivity to the historical rootedness of concepts, and with that their evidential sources of legitimacy and, above all, revitalization. Husserl is, after all, committed to the renewal of rationalism; and inwardness is invoked precisely in the context of a call to renewal.[7] Yet, as was already indicated above in connection with Husserl's concept of *Einströmen*, this sensitivity need not be limited to being the basis for transferring the results of transcendental investigation to the conceptual apparatus of a science such as psychology or even a philosophy of science. It could also have the function of highlighting historical possibilities that are inherent to the historicity as well as traditionality of the lifeworld itself, possibilities having to do with making decisions about the meaning or meaninglessness of things, or decisions on the level of what was above called the "claim of science." The origin of this new look of things would be the philosophical transformation of the meaning of subjectivity—and it is above all a new sense of self, as a rational subject, which finds its initial formulation in transcendental reflection and which is then "carried over" (again the theme of *Einströmen*) to natural life. The newfound sense of subjectivity works itself back into the natural rhythms of a life that, even if left behind in philosophical reflection, nevertheless "carries on"; though now its obviousnesses and traditionality are no longer the only sources of the sense of self, but form a complex whole with a new transcendental dimension that has found its "place" as an inwardness in concrete subjectivity.

But that does not yet tell us what "inwardness" amounts to. I would argue that it should not be understood as an "interiority," not even in the very sophisticated and profound manner in which the idea of subjective interiority is articulated in philosophies such as that of Michel Henry. If inwardness is understood as a kind of non-worldly event of the subjective, thus emphasizing its radical independence and, above all, invisibility, then the manner in which inwardness cleaves close to the world-opening experience of the understanding would be lost. For in Husserl, it is not an inwardness that belongs to life as such, to its being lived through, but an inwardness that specifically belongs to a rational subjectivity, or to the subjectivity of a life that has unfolded in the form of a rational understanding, which means a grasp of things and a projection of a world. The pursuit of the ideal of secure, "final" knowledge of the world as a world of things and persons, the historical task of rational life, is for Husserl the only dimension of life thanks to which a sensitivity of the sort represented by inwardness is possible at all. Philosophy must be prepared by a tradition that makes world-knowledge its object in a specific form; there are no exceptions to this. The same is true of inwardness: for inwardness is, at least on one level, the sensitivity to how and when concepts are being employed for the advance of an understanding, or a sensitivity to when we rely on an idea and when we do not. It also represents an awareness of the risks inherent to when we make an exception to a rule, or act outside the boundaries of what is familiar, or improvise on a theme. Such a sensitivity to the manner in which conceptuality—and, more generally, meaning—is woven throughout the patterns of lived experience is something that, for Husserl, is characteristic of a culture and a civilization that has been shaped by the tasks of science, and ultimately animated by a belief in its claim.

Whether it is true that a sensitivity of this type is possible only in a scientific culture is a question I would like to leave open. But I would like to stress the thesis that, whether it is scientific or not, it is only relative to a world-understanding that an inwardness can develop—which means that it cannot be interpreted as a pure homogenous interiority, one that is not developed on the basis of the patterns and style of world-understanding, but discovered as a hidden, invisible resource of subjective life. Inwardness is, as it were, the development of a reflected life, not reflected in a concept, but life as reflected in its own sensitivity to its overall style as a developing life of a "subject." Nevertheless, it is important to remember that for Husserl, this reflectivity of life is the result of an establishment (*Stiftung*) of reason in a particular form; it is not only a kind of thoughtfulness that is injected into rational understanding, but its reformation in light of its own inner idea, or telos.

In lectures and manuscripts of the later period, Husserl often invokes a Kantian conception of the moral subject, or the figure of a subject being held fast to a principle, or law, in accordance with which it fashions a sense for, and itself out of, moral autonomy. A Kantian inspiration is also behind the prominent theme of responsibility in the texts from the *Crisis* period. Yet for Husserl, the responsibility of philosophy is not only the phenomenon of speaking on behalf of that which is compelling, as was explored above, but more: it is the responsibility to live a certain way, in light of the apodeictic evidences of reason, for which one is responsible, not only with respect to their articulation, but for their employment in the determination of action and the formation of the shape of the surrounding cultural world. Yet this is really only a gesture. The notion of a transcendental inwardness is not enough to fulfill the promise of demonstrating how it is that pure reason, to use Kant's formulation of the problem, can be "practical." There are no incentives of pure practical reason in the sensitivity of inwardness, or anything in Husserl's transcendental philosophy as a whole that could assume the role played by respect in Kant's moral philosophy. All motivations in Husserl are historically conditioned, including rational motivations. Still, this gesture is important to understanding Husserl's ultimate philosophical goals: inwardness, as the result or impact of transcendental philosophy, is intended to represent the embodiment of a self understanding determined by the idea of a new humanity that, fashioned in a transcendental reflection on the sense and meaning of subjectivity, is to serve as a basis for the practicality of pure reason.

Yet my purpose above was not only to assess Husserl's self-interpretation, but to read Husserl's *Crisis* in such a way as to open a discussion concerning the potential for this kind of philosophy. This interpretation is, after all, being written in a very different time and in a very different philosophical climate than was the case when Husserl was working on the *Crisis*. The question with respect to the meaning of science is still very much relevant, I would argue, though it has become more and more infused with the question not so much of science, but of technology, which has in many ways become a new center of gravity with respect to the question of the nature of reason. The technologization of reason is of course an important theme in Husserl's later writings, but the question of technology today, it seems to me, has become not so much an issue of technologizing per se, but one of engineering. The complex and difficult legacy of the 20th century can be represented as both the task and the technological capacity to re-engineer the very look and feel of human experience itself. Just how such a task should be conceived, and how it should be pursued, is of course a very contentious issue; but that the technological capacity to radically reshape what it is to be a human being can never be without relevance for our future is beyond doubt. Luddism is dead.

Inwardness, I would argue, is one such latent potential of Husserlian thought, at least once certain aspects of it have been freed from the attempt to establish the foundation for a commitment to reason, or the responsibilities of philosophy. That is, once inwardness is not seen as the basis for the self awareness of a responsible subject in the Kantian sense, where the subject is determined by a clash between a call to responsibility and natural interests, but actually the contrary: inwardness as an awareness of a selfhood that is never absolutely, completely absorbed by its commitments, its valuations, its understanding, even the consciousness to which it belongs. This re-thinking of inwardness actually has a basis in Husserl's own thought, I would argue, as something already suggested by the concept of the lifeworld itself—or the lifeworld as "subjective" in the sense of a reserve that can never be fully articulated, or completely exhausted by understanding, whether it take the form of an rationalist ideology such as objectivism or any other form.

What, then, would such an inwardness be? It would not necessarily be an absolute self-possession of a subject that would serve as a foundation for a world-understanding; but it would be a dimension of subjective life thanks to which understanding would be lived in a self-conscious manner. It would not be a self-consciousness in the form of a self-objectification, whether as an idea of humanity or of a philosophical subject understood in a certain way—it would not be an identity; but it would be "personal," or a shape that personality itself could take, though "personality" in a very different sense from the Kantian *Persönlichkeit*. It would be, instead, inherently unstable, but not in danger of losing itself, or being lost; it would be the state of the subject which, after having found a means to make itself the subject of reflection thanks to phenomenology, finds itself "back" in life, struggling to find the words to articulate its own significance.

It would be, in other words, the very condition of the subject of scientific consciousness as it is presented in Part I of the *Crisis*, but now as one who is no longer haunted by the specter of skepticism, or the lack of faith—for there would no longer be a question of a faith in reason. There would be, instead, a capacity to ask different questions.

Notes

[1] Hua XXIX 343:5-7: "Jedes Erlebnis, jeder Akt, auch jeder Erkenntnisakt, und korrelativ, jede Erkenntnis (das Gebilde) ist motiviert, hat also seine Tradition."
[2] See Ideas I/Hua III §§77-78.
[3] See Hua VI §43; Hua XXIX 424-426.
[4] See Hua XXIX 77-83 (Text Nr. 7, "*Einströmen* (Sommer 1935). Cf. Hua XXIX 73:25-29; 136:5-30; 208:24-31.
[5] Hua XXIX 79:15-39.
[6] E.g. Hua XXIX 268:30-37: "Nach der Etablierung des philosophischen Lebens verbleibt die Fortsetzung des natürlichen als Rückkehr, nur daß dann das natürliche unter Erhaltung seiner alten Sinnes- und Seinsstruktur eine Verwandlung erfährt, daß <das> Ich, der Philosophierende, (in all <den> Mitphilosophierenden) durch Einströmen der transzendentalen Ergebnisse in die früher naiv apperzpierte Subjektivität eine transzendentale 'Innerlichkeit' erhält."
[7] See in particular Hua XXIX:6-10.

References

Husserl's Works

Husserliana, with English Translations

Ideen zu einer reinen Phänomenologie und phänomenologischen Philosophie. Erstes Buch: Allgemeine Einführung in die reine Phänomenologie, ed. Karl Schuhmann, Husserliana III,1. Hague 1976.

Ideas Pertaining to a Pure Phenomenology and a Phenomenological Philosophy. First Book: General Introduction to a Pure Phenomenology, trans. Fred Kersten, Husserl: Collected Works II. Hague 1982.

Ideen zu einer reinen Phänomenologie und phänomenologischen Philosophie. Zweites Buch: Phänomenologische Untersuchungen zur Konstitution, ed. Marly Biemel, Husserliana IV. Hague 1952.

Ideas Pertaining to a Pure Phenomenology and a Phenomenological Philosophy. Second Book: Studies in the Phenomenology of Constitution, trans. Richard Rojcewicz and André Schuwer, Collected Works III. Dordrecht 1989.

Die Krisis der europäischen Wissenschaften und transzendentale Phänomenologie, ed. Walter Biemel, Husserliana VI. Hague 1954.

The Crisis of European Sciences and Transcendental Phenomenology, trans. David Carr. Evanston 1970.

Erste Philosophie (1923-1924). Erster Teil: Kritische Ideengeschichte, ed. Rudolf Boehm, Husserliana VII. Hague 1956.

Erste Philosophie (1923-1924). Zweiter Teil: Theorie der phänomenologischen Reduktion, ed. Rudolf Boehm, Husserliana VIII. Hague 1959.

Phänomenologische Psychologie. Vorlesungen Sommersemester 1925, ed. Walter Biemel, Husserliana IX. Hague 1962.

Zur Phänomenologie des inneren Zeitbewusstseins (1893-1917), ed. Rudolf Boehm, Husserliana X. Hague 1966.

On the Phenomenology of the Consciousness of Internal Time (1893-1917). trans. John Barnett Brough, Edmund Husserl: Collected Works IV. Dordrecht 1991.

Analysen zur passiven Synthesis, ed. Margot Fleischer, Husserliana XI. Hague 1966.

Formale und transzendentale Logik. Versuch einer Kritik der logischen Vernunft. Mit ergänzenden Texten, ed. Paul Janssen, Husserliana XVII. Hague 1974.

Formal and Transcendental Logic, trans. Dorion Cairns. Hague 1969.

Logische Untersuchungen. Erster Band: Prolegomena zur reinen Logik, ed. Elmar Holenstein, Husserliana XVIII. Hague 1975.

Logische Untersuchungen. Zweiter Band: Untersuchungen zur Phänomenologie und Theorie der Erkenntnis, ed. Ursula Panzer, Husserliana XIX,1 and XIX,2. Hague 1984.

Logical Investigations. 2 volumes, trans. John Findlay. London 1970.

Aufsätze und Vorträge (1911-1921), ed. Thomas Nenon and Hans-Rainer Sepp, Husserliana XXV. Dordrecht 1986.

Husserl: Shorter Works, ed. Peter McCormick and Frederick Elliston. Notre Dame 1981.

Aufsätze und Vorträge (1922-1937) ed. Thomas Nenon and Hans-Rainer Sepp, Husserliana XXVII. Dordrecht 1989.

Die Krisis der europäischen Wissenschaften und die transzendentale Phänomenologie. Ergänzungsband. Texte aus dem Nachlaß 1934-1937, ed. Reinhold Smid, Husserliana XXIX. Dordrecht 1993.

Aktive Synthesen: Aus der Vorlesung "Transzendentale Logik" 1920-1921. Ergänzungsband zu "Analysen zur passiven Synthesis," ed. Roland Breeur, Husserliana XXXI. Dordrecht 2000.

Other than Husserliana

Husserl-Chronik. Denk- und Lebensweg Edmung Husserls, ed. Karl Schuhmann, Husserliana Dokumente I. Hague 1977.

Erfahrung und Urteil (1939), ed. Ludwig Landgrebe. Hamburg 1999.

Experience and Judgment, trans. James Churchill and Karl Ameriks, with an Afterword by Lothar Eley. Evanston 1973.

Literature

Aguirre, Antonio. Genetische Phänomenologie und Reduktion. Zur Letztbegründung der Wissenschaft aus der radikalen Skepsis im Denken E. Husserls, Phaenomenologica 38. Hague 1970.

Aristotle. Politics. Nicomachean Ethics. Metaphysics. Harvard, Loeb Classical Library.

St. Augustine. Confessions. Trans. F.J. Sheed. Indianapolis/Cambridge 1992.

Avenarius. Der menschliche Weltbegriff. Leipzig 1891.

________. Philosophie als Denken der Welt gemäß dem Prinzip des kleinsten Kraftmasses. Prolegomena zu einer Kritik der reinen Erfahrung. Leipzig 1876.

Bachelard, Suzanne. La Logique de Husserl. Étude zur Logique formelle et logique transcendentale. Paris 1957.

Banfi, A. Husserl et la crise de la pensée européenne, in: Husserl. Cahiers de Royaumont III, Paris 1959: 411-427.

Benoist, Jocelyn. Autour de Husserl. Paris 1994.

Bernet, Rudolf, Iso Kern, Eduard Marbach. Edmund Husserl: Darstellung seines Denkens. Hamburg 1989.

Blumenburg, Hans. Lebenswelt und Technisierung unter Aspekten der Phänomenologie, in: Filosofia (Torino) 14 (1963) 855-884.

________. Die Legitimität der Neuzeit. Frankfurt 1976.

Bochenski, J.M. Formale Logik, Orbis Academicus. Freiburg/München 1956.

Boehm, Rudolf. Vom Gesichtspunkt der Phänomenologie. Husserl-Studien. Hague 1968.

Brand, Gerd. Welt, Ich, und Zeit: Nach unveröffentlichen Manuskripten Edmund Husserls, Phaenomenologica. Hague 1955.

________. Die Lebenswelt. Eine Philosophie des konkreten Apriori. Berlin 1971.

Brentano, Franz. Psychologie vom empirischen Standpunkt, ed. Oskar Kraus. Hamburg 1955.

Cahoone, Lawrence. The Interpretation of Galilean Science: Cassirer Contrasted with Husserl and Heidegger, in: Studies in History and Philosophy of Science 17/1 (1986) 1-21.

Carr, David. Phenomenology and the Problem of History. Evanston 1974.

________ . Interpreting Husserl. Critical and Comparative Studies, Phaenomenologica 106. Hague 1987.

Cassirer, Ernst. Galileo's Platonism, in: Studies and Essays in the History of Science and Learning, ed. Montagu. New York 1944.

________. Wahrheitsbegriff und Wahrheitsproblem bei Galilei, in: Scientia (Milano) 62 (1937) 120-130, 185-193.

Cavaillès, Jean. Sur la logique et la théorie de la science, ed. Canguilhem and Ehresmann. Paris 1947.

Clavelin, Maurice. La philosophie naturelle de Galilée. Paris 1968.

Deleuze, Gilles. Différence et répétition. Paris 1968.
Depraz, Natalie. Temporalité et affection dans les manuscrits tardifs sur la temporalité, in: Temporalité et affection. Alter 2 (1994): 63-86.
________. Phenomenological Reduction and the Political, in: Husserl-Studies 12 (1995): 1-17.
Derrida, Jacques. Edmund Husserl's Origin of Geometry: An Introduction, trans. John Leavy. Lincoln 1989.
________. Le problème de la genèse dans la philosophie de Husserl. Paris 1990.
Descartes, René. Discours de la méthode, texte et commentaire par Étienne Gilson. Paris 1925.
De Warren, Nicolas. Philosophy and Human Perfection in the Cartesian Renaissance and its Modern Oblivion, in: Graduate Faculty Philosophy Journal 22/2 (2001) 185-212.
________. The Promise of Time: Time-Consciousness and the Breakthrough of Phenomenology. Dissertation. Boston University, 2001.
Dilthey, Wilhelm. Einleitung in die Geisteswissenschaften (1883), in: Gesammelte Schriften I. Stuttgart 1959.
Drake, Stillman. Galileo at Work: His Scientific Biography. Chicago 1978.
Dufrenne, Mikel. The Notion of the Apriori, trans. Edward Casey. Evanston 1966.
Faye, Emmanuel. Philosophie et Perfection de l'Homme: De la Renaissance à Descartes. Paris 1998.
Fein, Hubert. Genesis und Geltung in E. Husserls Phänomenologie, Kritische Studien zur Philosophie. Frankfurt/Wien 1970.
Gadamer, Hans-Georg. Neuere Philosophie I: Hegel. Husserl. Heidegger. Gesammelte Werke 3. Tübingen 1987.
Garrison, James. Husserl, Galileo, and the Processes of Idealization, in: Synthese 66 (1986) 329-338.
Gaukroger, Stephen. Explanatory Structures. Hassosks 1978.
Geymonat, Ludovico. Galileo Galilei, trans. Stillman Drake. New York 1965.
Gurwitsch, Aron. The Last Work of Edmund Husserl, in: Philosophy and Phenomenological Research XVI (1956) and XVII (1957).
________. Phenomenology and the Theory of Science, ed. Lester Embree. Evanston 1974.
Harvey, Charles and Jaako Hintikka. Modalization and Modalities, in: Phenomenology and the Formal Sciences, ed. John Drummond. Dordrecht 1991.
Heffernan, George. Bedeutung und Evidenz bei Edmund Husserl. Bonn 1983.

________. Am Anfang war die Logik. Amsterdam 1988.

________.Isagoge in die phänomenologische Apophantik, Phaenomenologica 107. Dordrecht 1989.

Heidegger, Martin. History of the Concept of Time: Prolegomena, trans. Theodor Kisiel. Bloomington 1992.

________. Sein und Zeit. Tübingen 1960.

________. Was ist Metaphysik? Frankfurt a.M. 1965.

________. Einführung in die Metaphysik. Tübingen 1966.

________. Die Zeit des Weltbildes (1938), in: Holzwege. Frankfurt a. M. 1950.

Held, Klaus. Lebendige Gegenwart. Die Frage nach der Seinsweise des transzendentalen Ich bei Edmund Husserl, entwickelt am Leitfaden der Zeitproblematik, Phaenomenologica 23. Hague 1966.

Henrich, Dieter. Kant's Notion of a Deduction and the Methodological Background of the First Critique, in: Kant's Transcendental Deductions, ed. E. Förster. Stanford 1989: 29-46.

Hintikka, Jaakko. Models for Modalities. Selected Essays. Dordrecht 1969.

Hohl, Hubert. Lebenswelt und Geschichte. Grundzüge der Spätphilosophie Husserls, Symposion. Freiburg/München 1962.

Holenstein, Elmar. Phänomenologie der Assoziation. Zur Struktur und Funktion eines Grundprinzips der passiven Genesis bei Edmund Husserl, Phaenomenologica. Hague 1972.

Hülsmann, Heinz. Zur Theorie der Sprache bei Edmund Husserl. München 1964.

Hylton, Peter. Russell, Idealism and the Emergence of Analytic Philosophy. Oxford 1990.

Janssen, Paul. Geschichte und Lebenswelt. Ein Beitrag zur Diskussion von Husserls Spätwerk, Phaenomenologica 35. Hague 1970.

________. Grundlagen der wissenschaftlichen Welterkenntnis. Frankfurt a. M. 1977.

Klein, Jakob. Die griechische Logistik und die Entstehung der Algebra, in: Quellen und Studien zur Geschichte der Mathematik, Astronomie und Physik, Abteilung B: Studien 3/1 (1934):18-105, 3/2 (1936):122-235.

Koselleck, Reinhardt. Kritik und Krise. Frankfurt a.M.: Suhrkamp 1973.

________. Article 'Krise,' in: Historisches Wörterbuch der Philosophie, ed. Joachim Ritter.

Koyré, Alexandre. Galileo Studies (1940), trans. J. Mepham. New Jersey 1978.

________. Galileo and Plato, in: Journal of the History of Ideas 4/4 (1943).

Landgrebe, Ludwig. Faktizität und Individuation. Studien zur Grundfragen der Phänomenologie, Hamburg 1982.

________. Husserls Abschied vom Cartesianismus, in: Der Weg der Phänomenologie. Das Problem einer ursprünglichen Erfahrung. Gütersloh: 1963: 163-206.

Leao, Maria. Art, Language, and Thought: Meaning and Expression in Hegel's Philosophy of Spirit. Dissertation: Boston University, 1996.

Lohmar, Dieter. Edmund Husserls >Formale und transzendentale Logik<. Darmstadt 2000.

________. Erfahrung und kategoriales Denken. Hume, Kant und Husserl über vorprädikative Erfahrung und prädikative Erkenntnis, Phaenomenologica 147. Dordrecht 1998.

________. Zu der Entstehung und den Ausgangsmaterialien von Edmund Husserl's Werk Erfahrung und Urteil, in: Husserl Studies 13 (1996): 31-71.

Lübbe, Hermann. Husserl und die europäische Krisis, in: Kant-Studien 49/1 (1957/8): 334-335.

Lucretius. The Nature of Things, trans. Frank Copley. New York 1977.

McCumber, John. The Company of Words: Hegel, Language, and Systematic Philosophy. Evanston 1993.

Merleau-Ponty, Maurice. Phénoménologie de la perception. Paris 1945.

________. Le visible et l'invisible. Paris 1964.

Meschkowski, Herbert. Problemgeschichte der Mathematik III. Mannheim 1986.

Montavont, Anne. De la passivité dans la phénoménologie de Husserl. Paris 1999.

Muralt, André de. L'Idee de la phénoménologie. L'exemplarisme husserlien. Paris 1958.

Nietzsche, Friedrich. Götzen-Dämmerung, in: Kritische Studienausgabe 6, ed. Giorgio Colli and Mazzino Montinari. München 1999.

Paci, Enzo. The Function of the Sciences and the Meaning of Man, trans. Evanston 1972.

Patočka, Jan. An Introduction to Husserl's Phenomenology, trans. Erazim Kohák, ed. James Dodd. Chicago 1996.

________ . Heretical Essays in the Philosophy of History, trans. Erazim Kohák. Chicago 1996.

________. Texte Dokumente Bibliographie. Hrsg. Ludger Hagedorn und Hans Rainer Sepp. Quellen: Orbis phaenomenologicus. Freiburg/München 1999.

Plato. Phaedrus. (Oxford)

Rickert, Heinrich. Die Grenzen der Naturwissenschaftlichen Begriffsbildung. Eine logische Einleitung in die historischen Wissenschaften. Tübingen 1929 (5th edition).

Ricoeur, Paul. Husserl: An Analysis of his Phenomenology, trans. Edward Ballard and Lester Embree. Evanston: Northwestern 1967.

Scheler, Max. Der Formalismus in der Ethik und die materiale Wertlehre. Bern/München 1966.

Schelling, F.W.J. Philosophie der Kunst, Schelling's Werke III. München 1927.

________ . System des transzendentalen Idealismus (1800). Hamburg: 1957.

Schiller, Friedrich. Was kann eine gute stehende Schaubühne eigentlich wirken? in: Schiller's Werke, Nationalausgabe 20. Weimar 1962.

Shea, William. Galileo's Intellectual Revolution. New York 1972.

Simmel, Georg. Die Probleme der Geschichtsphilosophie (1892), in: Georg Simmel Gesamtausgabe. Frankfurt a. M. 1989.

Sprute, J. Der Begriff Doxa in der platonischen Philosophie, Hypomnemata 2. Göttingen 1962.

Steinbock, Anthony. Home and Beyond: Generative Phenomenology After Husserl. Evanston 1995.

Ströker, Elizabeth. Geschichte und Lebenswelt als Sinnesfundament der Wissenschaften in Husserls Spätwerk, in: Lebenswelt und Wissenschaft in der Philosophie Edmund Husserl's, ed. E. Ströker. Frankfurt a.M.: 1979.

Tani, Toru. Life and the Lifeworld, in: Husserl-Studies 3 (1986): 57-78.

Thurnher, Rainer. Husserl's Idee einer 'wirklichen' echten Wissenschaftstheorie, in: Krise der Wissenschaften—Wissenschaft der Krisis?, ed. Helmuth Vetter. Wien 1998.

Weischedel, Wilhelm. Das Wesen der Verantwortung. Frankfurt a. M.: 1933.

Welton, Donn. The Origins of Meaning: A Critical Study of the Thresholds of Husserlian Phenomenology, Phaenomenologica. Hague 1983.

Yamaguchi, I. Passive Synthesis und Intersubjektivität bie Edmund Husserl. Hague 1982.

Index

Phaenomenologica

1. E. Fink: *Sein, Wahrheit, Welt.* Vor-Fragen zum Problem des Phänomen-Begriffs. 1958 ISBN 90-247-0234-8
2. H.L. van Breda and J. Taminiaux (eds.): *Husserl et la pensée moderne / Husserl und das Denken der Neuzeit.* Actes du deuxième Colloque International de Phénoménologie / Akten des zweiten Internationalen Phänomenologischen Kolloquiums (Krefeld, 1.–3. Nov. 1956). 1959 ISBN 90-247-0235-6
3. J.-C. Piguet: *De l'esthétique à la métaphysique.* 1960 ISBN 90-247-0236-4
4. *E. Husserl: 1850–1959.* Recueil commémoratif publié à l'occasion du centenaire de la naissance du philosophe. 1959 ISBN 90-247-0237-2

5/6. H. Spiegelberg: *The Phenomenological Movement.* A Historical Introduction. 3rd revised ed. with the collaboration of Karl Schuhmann. 1982 ISBN Hb: 90-247-2577-1; Pb: 90-247-2535-6

7. A. Roth: *Edmund Husserls ethische Untersuchungen.* Dargestellt anhand seiner Vorlesungsmanuskripte. 1960 ISBN 90-247-0241-0
8. E. Levinas: *Totalité et infini.* Essai sur l'extériorité. 4th ed., 4th printing 1984 ISBN Hb: 90-247-5105-5; Pb: 90-247-2971-8
9. A. de Waelhens: *La philosophie et les expériences naturelles.* 1961 ISBN 90-247-0243-7
10. L. Eley: *Die Krise des Apriori in der transzendentalen Phänomenologie Edmund Husserls.* 1962 ISBN 90-247-0244-5
11. A. Schutz: *Collected Papers, I.* The Problem of Social Reality. Edited and introduced by M. Natanson. 1962; 5th printing: 1982 ISBN Hb: 90-247-5089-X; Pb: 90-247-3046-5
    Collected Papers, II *see* below under Volume 15
    Collected Papers, III *see* below under Volume 22
    Collected Papers, IV *see* below under Volume 136
12. J.M. Broekman: *Phänomenologie und Egologie.* Faktisches und transzendentales Ego bei Edmund Husserl. 1963 ISBN 90-247-0245-3
13. W.J. Richardson: *Heidegger. Through Phenomenology to Thought.* Preface by Martin Heidegger. 1963; 3rd printing: 1974 ISBN 90-247-02461-1
14. J.N. Mohanty: *Edmund Husserl's Theory of Meaning.* 1964; reprint: 1969 ISBN 90-247-0247-X
15. A. Schutz: *Collected Papers, II.* Studies in Social Theory. Edited and introduced by A. Brodersen. 1964; reprint: 1977 ISBN 90-247-0248-8
16. I. Kern: *Husserl und Kant.* Eine Untersuchung über Husserls Verhältnis zu Kant und zum Neukantianismus. 1964; reprint: 1984 ISBN 90-247-0249-6
17. R.M. Zaner: *The Problem of Embodiment.* Some Contributions to a Phenomenology of the Body. 1964; reprint: 1971 ISBN 90-247-5093-8
18. R. Sokolowski: *The Formation of Husserl's Concept of Constitution.* 1964; reprint: 1970 ISBN 90-247-5086-5
19. U. Claesges: *Edmund Husserls Theorie der Raumkonstitution.* 1964 ISBN 90-247-0251-8
20. M. Dufrenne: *Jalons.* 1966 ISBN 90-247-0252-6
21. E. Fink: *Studien zur Phänomenologie, 1930–1939.* 1966 ISBN 90-247-0253-4
22. A. Schutz: *Collected Papers, III.* Studies in Phenomenological Philosophy. Edited by I. Schutz. With an introduction by Aron Gurwitsch. 1966; reprint: 1975 ISBN 90-247-5090-3
23. K. Held: *Lebendige Gegenwart.* Die Frage nach der Seinsweise des transzendentalen Ich bei Edmund Husserl, entwickelt am Leitfaden der Zeitproblematik. 1966 ISBN 90-247-0254-2
24. O. Laffoucrière: *Le destin de la pensée et 'La Mort de Dieu' selon Heidegger.* 1968 ISBN 90-247-0255-0
25. E. Husserl: *Briefe an Roman Ingarden.* Mit Erläuterungen und Erinnerungen an Husserl. Hrsg. von R. Ingarden. 1968 ISBN Hb: 90-247-0257-7; Pb: 90-247-0256-9
26. R. Boehm: *Vom Gesichtspunkt der Phänomenologie* (I). Husserl-Studien. 1968 ISBN Hb: 90-247-0259-3; Pb: 90-247-0258-5
    For *Band II* see below under Volume 83

Phaenomenologica

27. T. Conrad: *Zur Wesenslehre des psychischen Lebens und Erlebens.* Mit einem Geleitwort von H.L. van Breda. 1968 ISBN 90-247-0260-7
28. W. Biemel: *Philosophische Analysen zur Kunst der Gegenwart.* 1969 ISBN Hb: 90-247-0263-1; Pb: 90-247-0262-3
29. G. Thinès: *La problématique de la psychologie.* 1968 ISBN Hb: 90-247-0265-8; Pb: 90-247-0264-X
30. D. Sinha: *Studies in Phenomenology.* 1969 ISBN Hb: 90-247-0267-4; Pb: 90-247-0266-6
31. L. Eley: *Metakritik der formalen Logik.* Sinnliche Gewissheit als Horizont der Aussagenlogik und elementaren Prädikatenlogik. 1969 ISBN Hb: 90-247-0269-0; Pb: 90-247-0268-2
32. M.S. Frings: *Person und Dasein.* Zur Frage der Ontologie des Wertseins. 1969 ISBN Hb: 90-247-0271-2; Pb: 90-247-0270-4
33. A. Rosales: *Transzendenz und Differenz.* Ein Beitrag zum Problem der ontologischen Differenz beim frühen Heidegger. 1970 ISBN 90-247-0272-0
34. M.M. Saraiva: *L'imagination selon Husserl.* 1970 ISBN 90-247-0273-9
35. P. Janssen: *Geschichte und Lebenswelt.* Ein Beitrag zur Diskussion von Husserls Spätwerk. 1970 ISBN 90-247-0274-7
36. W. Marx: *Vernunft und Welt.* Zwischen Tradition und anderem Anfang. 1970 ISBN 90-247-5042-3
37. J.N. Mohanty: *Phenomenology and Ontology.* 1970 ISBN 90-247-5053-9
38. A. Aguirre: *Genetische Phänomenologie und Reduktion.* Zur Letztbegründung der Wissenschaft aus der radikalen Skepsis im Denken E. Husserls. 1970 ISBN 90-247-5025-3
39. T.F. Geraets: *Vers une nouvelle philosophie transcendentale.* La genèse de la philosophie de Maurice Merleau-Ponty jusqu'à la 'Phénoménologie de la perception.' Préface par E. Levinas. 1971 ISBN 90-247-5024-5
40. H. Declève: *Heidegger et Kant.* 1970 ISBN 90-247-5016-4
41. B. Waldenfels: *Das Zwischenreich des Dialogs.* Sozialphilosophische Untersuchungen in Anschluss an Edmund Husserl. 1971 ISBN 90-247-5072-5
42. K. Schuhmann: *Die Fundamentalbetrachtung der Phänomenologie.* Zum Weltproblem in der Philosophie Edmund Husserls. 1971 ISBN 90-247-5121-7
43. K. Goldstein: *Selected Papers/Ausgewählte Schriften.* Edited by A. Gurwitsch, E.M. Goldstein Haudek and W.E. Haudek. Introduction by A. Gurwitsch. 1971 ISBN 90-247-5047-4
44. E. Holenstein: *Phänomenologie der Assoziation.* Zu Struktur und Funktion eines Grundprinzips der passiven Genesis bei E. Husserl. 1972 ISBN 90-247-1175-4
45. F. Hammer: *Theonome Anthropologie?* Max Schelers Menschenbild und seine Grenzen. 1972 ISBN 90-247-1186-X
46. A. Pažanin: *Wissenschaft und Geschichte in der Phänomenologie Edmund Husserls.* 1972 ISBN 90-247-1194-0
47. G.A. de Almeida: *Sinn und Inhalt in der genetischen Phänomenologie E. Husserls.* 1972 ISBN 90-247-1318-8
48. J. Rolland de Renéville: *Aventure de l'absolu.* 1972 ISBN 90-247-1319-6
49. U. Claesges und K. Held (eds.): *Perspektiven transzendental-phänomenologischer Forschung.* Für Ludwig Landgrebe zum 70. Geburtstag von seiner Kölner Schülern. 1972 ISBN 90-247-1313-7
50. F. Kersten and R. Zaner (eds.): *Phenomenology: Continuation and Criticism.* Essays in Memory of Dorion Cairns. 1973 ISBN 90-247-1302-1
51. W. Biemel (ed.): *Phänomenologie Heute.* Festschrift für Ludwig Landgrebe. 1972 ISBN 90-247-1336-6
52. D. Souche-Dagues: *Le développement de l'intentionalité dans la phénoménologie husserlienne.* 1972 ISBN 90-247-1354-4
53. B. Rang: *Kausalität und Motivation.* Untersuchungen zum Verhältnis von Perspektivität und Objektivität in der Phänomenologie Edmund Husserls. 1973 ISBN 90-247-1353-6
54. E. Levinas: *Autrement qu'être ou au-delà de l'essence.* 2nd. ed.: 1978 ISBN 90-247-2030-3
55. D. Cairns: *Guide for Translating Husserl.* 1973 ISBN Pb: 90-247-1452-4

Phaenomenologica

56. K. Schuhmann: *Die Dialektik der Phänomenologie, I.* Husserl über Pfänder. 1973 ISBN 90-247-1316-1
57. K. Schuhmann: *Die Dialektik der Phänomenologie, II.* Reine Phänomenologie und phänomenologische Philosophie. Historisch-analytische Monographie über Husserls 'Ideen I'. 1973 ISBN 90-247-1307-2
58. R. Williame: *Les fondements phénoménologiques de la sociologie compréhensive: Alfred Schutz et Max Weber.* 1973 ISBN 90-247-1531-8
59. E. Marbach: *Das Problem des Ich in der Phänomenologie Husserls.* 1974 ISBN 90-247-1587-3
60. R. Stevens: *James and Husserl: The Foundations of Meaning.* 1974 ISBN 90-247-1631-4
61. H.L. van Breda (ed.): *Vérité et Vérification / Wahrheit und Verifikation.* Actes du quatrième Colloque International de Phénoménologie / Akten des vierten Internationalen Kolloquiums für Phänomenologie (Schwäbisch Hall, Baden-Württemberg, 8.–11. September 1969). 1974 ISBN 90-247-1702-7
62. Ph.J. Bossert (ed.): *Phenomenological Perspectives.* Historical and Systematic Essays in Honor of Herbert Spiegelberg. 1975. ISBN 90-247-1701-9
63. H. Spiegelberg: *Doing Phenomenology.* Essays on and in Phenomenology. 1975 ISBN 90-247-1725-6
64. R. Ingarden: *On the Motives which Led Husserl to Transcendental Idealism.* 1975 ISBN 90-247-1751-5
65. H. Kuhn, E. Avé-Lallemant and R. Gladiator (eds.): *Die Münchener Phänomenologie.* Vorträge des Internationalen Kongresses in München (13.–18. April 1971). 1975 ISBN 90-247-1740-X
66. D. Cairns: *Conversations with Husserl and Fink.* Edited by the Husserl-Archives in Louvain. With a foreword by R.M. Zaner. 1975 ISBN 90-247-1793-0
67. G. Hoyos Vásquez: *Intentionalität als Verantwortung.* Geschichtsteleologie und Teleologie der Intentionalität bei Husserl. 1976 ISBN 90-247-1794-9
68. J. Patočka: *Le monde naturel comme problème philosophique.* 1976 ISBN 90-247-1795-7
69. W.W. Fuchs: *Phenomenology and the Metaphysics of Presence.* An Essay in the Philosophy of Edmund Husserl. 1976 ISBN 90-247-1822-8
70. S. Cunningham: *Language and the Phenomenological Reductions of Edmund Husserl.* 1976 ISBN 90-247-1823-6
71. G.C. Moneta: *On Identity.* A Study in Genetic Phenomenology. 1976 ISBN 90-247-1860-0
72. W. Biemel und das Husserl-Archiv zu Löwen (eds.): *Die Welt des Menschen – Die Welt der Philosophie.* Festschrift für Jan Patočka. 1976 ISBN 90-247-1899-6
73. M. Richir: *Au-delà du renversement copernicien.* La question de la phénoménologie et son fondement. 1976 ISBN 90-247-1903-8
74. H. Mongis: *Heidegger et la critique de la notion de valeur.* La destruction de la fondation métaphysique. Lettre-préface de Martin Heidegger. 1976 ISBN 90-247-1904-6
75. J. Taminiaux: *Le regard et l'excédent.* 1977 ISBN 90-247-2028-1
76. Th. de Boer: *The Development of Husserl's Thought.* 1978 ISBN Hb: 90-247-2039-7; Pb: 90-247-2124-5
77. R.R. Cox: *Schutz's Theory of Relevance.* A Phenomenological Critique. 1978 ISBN 90-247-2041-9
78. S. Strasser: *Jenseits von Sein und Zeit.* Eine Einführung in Emmanuel Levinas' Philosophie. 1978 ISBN 90-247-2068-0
79. R.T. Murphy: *Hume and Husserl.* Towards Radical Subjectivism. 1980 ISBN 90-247-2172-5
80. H. Spiegelberg: *The Context of the Phenomenological Movement.* 1981 ISBN 90-247-2392-2
81. J.R. Mensch: *The Question of Being in Husserl's Logical Investigations.* 1981 ISBN 90-247-2413-9
82. J. Loscerbo: *Being and Technology.* A Study in the Philosophy of Martin Heidegger. 1981 ISBN 90-247-2411-2
83. R. Boehm: *Vom Gesichtspunkt der Phänomenologie II.* Studien zur Phänomenologie der Epoché. 1981 ISBN 90-247-2415-5

Phaenomenologica

84. H. Spiegelberg and E. Avé-Lallemant (eds.): *Pfänder-Studien.* 1982 ISBN 90-247-2490-2
85. S. Valdinoci: *Les fondements de la phénoménologie husserlienne.* 1982 ISBN 90-247-2504-6
86. I. Yamaguchi: *Passive Synthesis und Intersubjektivität bei Edmund Husserl.* 1982 ISBN 90-247-2505-4
87. J. Libertson: *Proximity.* Levinas, Blanchot, Bataille and Communication. 1982 ISBN 90-247-2506-2
88. D. Welton: *The Origins of Meaning.* A Critical Study of the Thresholds of Husserlian Phenomenology. 1983 ISBN 90-247-2618-2
89. W.R. McKenna: *Husserl's 'Introductions to Phenomenology'.* Interpretation and Critique. 1982 ISBN 90-247-2665-4
90. J.P. Miller: *Numbers in Presence and Absence.* A Study of Husserl's Philosophy of Mathematics. 1982 ISBN 90-247-2709-X
91. U. Melle: *Das Wahrnehmungsproblem und seine Verwandlung in phänomenologischer Einstellung.* Untersuchungen zu den phänomenologischen Wahrnehmungstheorien von Husserl, Gurwitsch und Merleau-Ponty. 1983 ISBN 90-247-2761-8
92. W.S. Hamrick (ed.): *Phenomenology in Practice and Theory.* Essays for Herbert Spiegelberg. 1984 Hb: ISBN 90-247-2926-2; Pb: 90-247-3197-6
93. H. Reiner: *Duty and Inclination.* The Fundamentals of Morality Discussed and Redefined with Special Regard to Kant and Schiller. 1983 ISBN 90-247-2818-5
94. M.J. Harney: *Intentionality, Sense and the Mind.* 1984 ISBN 90-247-2891-6
95. Kah Kyung Cho (ed.): *Philosophy and Science in Phenomenological Perspective.* 1984 ISBN 90-247-2922-X
96. A. Lingis: *Phenomenological Explanations.* 1986 ISBN Hb: 90-247-3332-4; Pb: 90-247-3333-2
97. N. Rotenstreich†: *Reflection and Action.* 1985 ISBN Hb: 90-247-2969-6; Pb: 90-247-3128-3
98. J.N. Mohanty: *The Possibility of Transcendental Philosophy.* 1985 ISBN Hb: 90-247-2991-2; Pb: 90-247-3146-1
99. J.J. Kockelmans: *Heidegger on Art and Art Works.* 1985 ISBN Hb: 90-247-3102-X; Pb ISBN 90-247-3144-5
100. E. Lévinas: *Collected Philosophical Papers.* 1987 ISBN Hb: 90-247-3272-7; Pb: 90-247-3395-2
101. R. Regvald: *Heidegger et le problème du néant.* 1986 ISBN 90-247-3388-X
102. J.A. Barash: *Martin Heidegger and the Problem of Historical Meaning.* 1987 ISBN 90-247-3493-2
103. J.J. Kockelmans (ed.): *Phenomenological Psychology.* The Dutch School. 1987 ISBN 90-247-3501-7
104. W.S. Hamrick: *An Existential Phenomenology of Law: Maurice Merleau-Ponty.* 1987 ISBN 90-247-3520-3
105. J.C. Sallis, G. Moneta and J. Taminiaux (eds.): *The Collegium Phaenomenologicum. The First Ten Years.* 1988 ISBN 90-247-3709-5
106. D. Carr: *Interpreting Husserl.* Critical and Comparative Studies. 1987. ISBN 90-247-3505-X
107. G. Heffernan: *Isagoge in die phänomenologische Apophantik.* Eine Einführung in die phänomenologische Urteilslogik durch die Auslegung des Textes der *Formalen und transzendentalen Logik* von Edmund Husserl. 1989 ISBN 90-247-3710-9
108. F. Volpi, J.-F. Mattéi, Th. Sheenan, J.-F. Courtine, J. Taminiaux, J. Sallis, D. Janicaud, A.L. Kelkel, R. Bernet, R. Brisart, K. Held, M. Haar et S. IJsseling: *Heidegger et l'idée de la phénoménologie.* 1988 ISBN 90-247-3586-6
109. C. Singevin: *Dramaturgie de l'esprit.* 1988 ISBN 90-247-3557-2
110. J. Patočka: *Le monde naturel et le mouvement de l'existence humaine.* 1988 ISBN 90-247-3577-7
111. K.-H. Lembeck: *Gegenstand Geschichte.* Geschichtswissenschaft in Husserls Phänomenologie. 1988 ISBN 90-247-3635-8
112. J.K. Cooper-Wiele: *The Totalizing Act.* Key to Husserl's Early Philosophy. 1989 ISBN 0-7923-0077-7

Phaenomenologica

113. S. Valdinoci: *Le principe d'existence.* Un devenir psychiatrique de la phénoménologie. 1989 ISBN 0-7923-0125-0
114. D. Lohmar: *Phänomenologie der Mathematik.* 1989 ISBN 0-7923-0187-0
115. S. IJsseling (Hrsgb.): *Husserl-Ausgabe und Husserl-Forschung.* 1990 ISBN 0-7923-0372-5
116. R. Cobb-Stevens: *Husserl and Analytic Philosophy.* 1990 ISBN 0-7923-0467-5
117. R. Klockenbusch: *Husserl und Cohn.* Widerspruch, Reflexion und Telos in Phänomenologie und Dialektik. 1989 ISBN 0-7923-0515-9
118. S. Vaitkus: *How is Society Possible?* Intersubjectivity and the Fiduciary Attitude as Problems of the Social Group in Mead, Gurwitsch, and Schutz. 1991 ISBN 0-7923-0820-4
119. C. Macann: *Presence and Coincidence.* The Transformation of Transcendental into Ontological Phenomenology. 1991 ISBN 0-7923-0923-5
120. G. Shpet: *Appearance and Sense.* Phenomenology as the Fundamental Science and Its Problems. Translated from Russian by Th. Nemeth. 1991 ISBN 0-7923-1098-5
121. B. Stevens: *L'apprentissage des signes.* Lecture de Paul Ricœur. 1991 ISBN 0-7923-1244-9
122. G. Soffer: *Husserl and the Question of Relativism.* 1991 ISBN 0-7923-1291-0
123. G. Römpp: *Husserls Phänomenologie der Intersubjektivität.* Und Ihre Bedeutung für eine Theorie intersubjektiver Objektivität und die Konzeption einer phänomenologischen Philosophie. 1991 ISBN 0-7923-1361-5
124. S. Strasser: *Welt im Widerspruch.* Gedanken zu einer Phänomenologie als ethischer Fundamentalphilosophie. 1991 ISBN Hb: 0-7923-1404-2; Pb: 0-7923-1551-0
125. R.P. Buckley: *Husserl, Heidegger and the Crisis of Philosophical Responsibility.* 1992 ISBN 0-7923-1633-9
126. J.G. Hart: *The Person and the Common Life.* Studies in a Husserlian Social Ethics. 1992 ISBN 0-7923-1724-6
127. P. van Tongeren, P. Sars, C. Bremmers and K. Boey (eds.): *Eros and Eris.* Contributions to a Hermeneutical Phenomenology. Liber Amicorum for Adriaan Peperzak. 1992 ISBN 0-7923-1917-6
128. Nam-In Lee: *Edmund Husserls Phänomenologie der Instinkte.* 1993 ISBN 0-7923-2041-7
129. P. Burke and J. Van der Veken (eds.): *Merleau-Ponty in Contemporary Perspective.* 1993 ISBN 0-7923-2142-1
130. G. Haefliger: *Über Existenz: Die Ontologie Roman Ingardens.* 1994 ISBN 0-7923-2227-4
131. J. Lampert: *Synthesis and Backward Reference in Husserl's* Logical Investigations. 1995 ISBN 0-7923-3105-2
132. J.M. DuBois: *Judgment and Sachverhalt.* An Introduction to Adolf Reinach's Phenomenological Realism. 1995 ISBN 0-7923-3519-8
133. B.E. Babich (ed.): *From Phenomenology to Thought, Errancy, and Desire.* Essays in Honor of William J. Richardson, S.J. 1995 ISBN 0-7923-3567-8
134. M. Dupuis: *Pronoms et visages.* Lecture d'Emmanuel Levinas. 1996 ISBN Hb: 0-7923-3655-0; Pb 0-7923-3994-0
135. D. Zahavi: *Husserl und die transzendentale Intersubjektivität.* Eine Antwort auf die sprachpragmatische Kritik. 1996 ISBN 0-7923-3713-1
136. A. Schutz: *Collected Papers, IV.* Edited with preface and notes by H. Wagner† and G. Psathas, in collaboration with F. Kersten. 1996 ISBN 0-7923-3760-3
137. P. Kontos: *D'une phénoménologie de la perception chez Heidegger.* 1996 ISBN 0-7923-3776-X
138. F. Kuster: *Wege der Verantwortung.* Husserls Phänomenologie als Gang durch die Faktizität. 1996 ISBN 0-7923-3916-9
139. C. Beyer: *Von Bolzano zu Husserl.* Eine Untersuchung über den Ursprung der phänomenologischen Bedeutungslehre. 1996 ISBN 0-7923-4050-7
140. J. Dodd: *Idealism and Corporeity.* An Essay on the Problem of the Body in Husserl's Phenomenology. 1997 ISBN 0-7923-4400-6
141. E. Kelly: *Structure and Diversity.* Studies in the Phenomenological Philosophy of Max Scheler. 1997 ISBN 0-7923-4492-8

Phaenomenologica

142. J. Cavallin: *Content and Object.* Husserl, Twardowski and Psychologism. 1997
ISBN 0-7923-4734-X
143. H.P. Steeves: *Founding Community.* A Phenomenological-Ethical Inquiry. 1997
ISBN 0-7923-4798-6
144. M. Sawicki: *Body, Text, and Science.* The Literacy of Investigative Practices and the Phenomenology of Edith Stein. 1997 ISBN 0-7923-4759-5; Pb: 1-4020-0262-9
145. O.K. Wiegand: *Interpretationen der Modallogik.* Ein Beitrag zur phänomenologischen Wissenschaftstheorie. 1998 ISBN 0-7923-4809-5
146. P. Marrati-Guénoun: *La genèse et la trace.* Derrida lecteur de Husserl et Heidegger. 1998
ISBN 0-7923-4969-5
147. D. Lohmar: *Erfahrung und kategoriales Denken.* Hume, Kant und Husserl über vorprädikative Erfahrung und prädikative Erkenntnis. 1998 ISBN 0-7923-5117-7
148. N. Depraz and D. Zahavi (eds.): *Alterity and Facticity.* New Perspectives on Husserl. 1998
ISBN 0-7923-5187-8
149. E. Øverenget: *Seeing the Self.* Heidegger on Subjectivity. 1998
ISBN Hb: 0-7923-5219-X; Pb: 1-4020-0259-9
150. R.D. Rollinger: *Husserls Position in the School of Brentano.* 1999 ISBN 0-7923-5684-5
151. A. Chrudzimski: *Die Erkenntnistheorie von Roman Ingarden.* 1999 ISBN 0-7923-5688-8
152. B. Bergo: *Levinas Between Ethics and Politics.* For the Beauty that Adorns the Earth. 1999
ISBN 0-7923-5694-2
153. L. Ni: *Seinsglaube in der Phänomenologie Edmund Husserls.* 1999 ISBN 0-7923-5779-5
154. E. Feron: *Phénoménologie de la mort.* Sur les traces de Levinas. 1999 ISBN 0-7923-5935-6
155. R. Visker: *Truth and Singularity.* Taking Foucault into Phenomenology. 1999
ISBN Hb: 0-7923-5985-2; Pb: 0-7923-6397-3
156. E.E. Kleist: *Judging Appearances.* A Phenomenological Study of the Kantian *sensus communis.* 2000 ISBN Hb: 0-7923-6310-8; Pb: 1-4020-0258-0
157. D. Pradelle: *L'archéologie du monde.* Constitution de l'espace, idéalisme et intuitionnisme chez Husserl. 2000 ISBN 0-7923-6313-2
158. H.B. Schmid: *Subjekt, System, Diskurs.* Edmund Husserls Begriff transzendentaler Subjektivität in sozialtheoretischen Bezügen. 2000 ISBN 0-7923-6424-4
159. A. Chrudzimski: *Intentionalitätstheorie beim frühen Brentano.* 2001 ISBN 0-7923-6860-6
160. N. Depraz: *Lucidité du corps.* De l'empirisme transcendantal en phénoménologie. 2001
ISBN 0-7923-6977-7
161. T. Kortooms: *Phenomenology of Time.* Edmund Husserl's Analysis of Time-Consciousness. 2001
ISBN 1-4020-0121-5
162. R. Boehm: *Topik.* 2002 ISBN 1-4020-0629-2
163. A. Chernyakov: *The Ontology of Time.* Being and Time in the Philosophies of Aristotle, Husserl and Heidegger. 2002 ISBN 1-4020-0682-9
164. D. Zahavi and F. Stjernfelt (eds.): *One Hundred Years of Phenomenology.* Husserl's *Logical Investigations* Revisited. 2002 ISBN 1-4020-0700-0
165. B. Ferreira: *Stimmung bei Heidegger.* Das Phänomen der Stimmung im Kontext von Heideggers Existenzialanalyse des Daseins. 2002 ISBN 1-4020-0701-9
166. S. Luft: *"Phänomenologie der Phänomenologie".* Systematik und Methodologie der Phänomenologie in der Auseinandersetzung zwischen Husserl und Fink. 2002
ISBN 1-4020-0901-1
167. M. Roesner: *Metaphysica ludens.* Das Spiel als phänomenologische Grundfigur im Denken Martin Heideggers. 2003 ISBN 1-4020-1234-9
168. B. Bouckaert: *L'idée de l'autre.* La question de l'idéalité et de l'altérité chez Husserl des *Logische Untersuchungen* aux *Ideen I.* 2003 ISBN 1-4020-1262-4
169. M.S. Frings: *LifeTime. Max Scheler's Philosophy of Time.* A First Inquiry and Presentation. 2003
ISBN 1-4020-1333-7

Phaenomenologica

170. T. Stähler: *Die Unruhe des Anfangs*. Hegel und Husserl über den Weg in die Phänomenologie. 2003 ISBN 1-4020-1547-X
171. P. Quesne: *Les* Recherches philosophiques *du jeune Heidegger*. 2003 ISBN 1-4020-1671-9
172. A. Chrudzimski: *Die Ontologie Franz Brentanos*. 2004 ISBN 1-4020-1859-2
173. S. Overgaard: *Husserl and Heidegger on Being in the World*. 2004 ISBN 1-4020-2043-0
174. J. Dodd: *Crisis and Reflection*. An Essay on Husserl's *Crisis of the European Sciences*. 2004 ISBN 1-4020-2174-7

Previous volumes are still available

Further information about *Phenomenology* publications is available on request

Kluwer Academic Publishers – Dordrecht / Boston / London

MIX
Papier aus verantwortungsvollen Quellen
Paper from responsible sources
FSC® C105338

If you have any concerns about our products,
you can contact us on
ProductSafety@springernature.com

In case Publisher is established outside the EU,
the EU authorized representative is:
Springer Nature Customer Service Center GmbH
Europaplatz 3, 69115 Heidelberg, Germany

Printed by Libri Plureos GmbH
in Hamburg, Germany